Jewish Intellectuals and the University

Jewish Intellectuals and the University

Marla Morris

JEWISH INTELLECTUALS AND THE UNIVERSITY

First published in 2006 by
PALGRAVE MACMILLAN™
175 Fifth Avenue, New York, N.Y. 10010 and
Houndmills, Basingstoke, Hampshire, England RG21 6XS
Companies and representatives throughout the world.

PALGRAVE MACMILLAN is the global academic imprint of the Palgrave Macmillan division of St. Martin's Press, LLC and of Palgrave Macmillan Ltd. Macmillan® is a registered trademark in the United States, United Kingdom and other countries. Palgrave is a registered trademark in the European Union and other countries.

ISBN-13: 978–1–4039–7580–5
ISBN-10: 1–4039–7580–9

Library of Congress Cataloging-in-Publication Data is available from the Library of Congress.

A catalogue record for this book is available from the British Library.

Design by Newgen Imaging Systems (P) Ltd., Chennai, India.

First edition: December 2006

10 9 8 7 6 5 4 3 2 1

Printed in the United States of America.

I would like to dedicate this book to my grandmother
Sally Stiller, the guiding light of my family.

Contents

Acknowledgments

I would like to thank Bill Pinar, Alan Block, Michael Eigen, Sander Gilman, Naomi Rucker, and especially Mary Aswell Doll. This book would not have been possible without the influence of these people.

PART I

Jewish Intellectuals

CHAPTER 1

Introduction: Tropes of Otherness and Jewish Intellectuals

This book is a study of Otherness as experienced by Jewish intellectuals who grapple with anti-Semitism and bureaucratic chaos within the halls of academe. Jewish intellectuals need to embrace the Otherness of the dystopic university. The dystopic university is a site that founders; it is a site that is confusing and schizophrenic; the dystopic university is an institution that oppresses yet can also be somewhat freeing. Despite this schizoid atmosphere it is still possible for Jewish scholars to find "lines of flight" out (Deleuze and Guattari, 1987, p. 32). The flight out of oppression begins with study and scholarship. Through scholarship one finds emotional and intellectual freedom.

The first part of this book examines Jewish intellectuals as a trope of Otherness. This book explores what the *experience* of Otherness *feels* like for Jews. Thus, this study is profoundly phenomenological, psychoanalytic, and historical. This book examines historically and culturally European Jewish intellectual life before the rise of the Third Reich. In my previous work (Morris, 2001), I grappled with the Third Reich and argued for a dystopic curriculum when studying the Holocaust. The question here turns on the ways in which Jews survived emotionally and intellectually during the late years of the Habsburg Empire up to the Holocaust. Jewish European culture and the experience of schooling are examined during the reign of Franz Joseph and the early twentieth century. There are certainly connections between European and American experiences of schooling. Schooling experiences for Jews—in Europe as well as in America—estrange.

The second part of the book turns toward the problem of subjectivity and interiority. Here the trope of madness (as an extreme state of interiority) is fleshed out. This chapter explores chaotic feeling states that may be understood metaphorically when applied to the

Jewish experience of being othered. Drawing on the work of Sander Gilman (1993), the argument here is not that Jews are inherently mad or pathological, as was the case in nineteenth century medical circles. Rather, the point here is that external pressures such as anti-Semitism and bureaucratic chaos—as experienced in universities or even madhouses—may evoke extreme feeling states. Drawing on Michael Eigen (2001a), the suggestion in this book is that one must "make room" (p. 5) for extreme chaotic feeling states—in order to survive. If one represses chaotic emotional states, the repressed returns in split off and unproductive ways. In order to be a productive scholar one must be a little mad. The reader is asked here to explore what it *feels* like to be mad. After studying madness, one might be better able to understand what it feels like to live in sites that are ungrounded and schizophrenic, like the university.

Finally, the third section of the book examines the trope of the dystopic university. The university is a schizoid place where scholars experience bewildering bureaucracy *and* the call to intellectual work; scholars experience both oppression and freedom. Jewish intellectuals must work around anti-Semitism and mindless bureaucracy to do their work. Working inside the dystopic university evokes feeling states that are uncertain, chaotic and, in a sense, mad. The situation for Jews—who are housed in the university—is especially precarious.

Dystopic means chaos. Dystopia means post–9-11. Here, the university has no aim. We are no longer certain what our task is as scholars. Scholars can use this chaos and madness to create intellectual ways out. We live in an age of terror. Never before has America lived in such a dystopic condition. Doomsday plans are in place for Congress. This seems the stuff of science fiction, but it is not. The dystopic university demands that we begin to explore chaotic states of being. The rubble of the Twin Towers is now part of our history and part of our psychic make-up.

From a phenomenological perspective, what is the experience of feeling at odds with one's surroundings, especially for Jewish scholars? This push and pull of university life creates feelings of groundlessness. The dystopic university is not a place of nihilism and cynicism. NO. That is not the point of this book. Doing intellectual work is, in a sense, freeing. It helps one better understand these chaotic times. The dystopic university creates spaces of Otherness where intellectuals might begin to explore what it means to reconstruct the public space of the academy in new and exciting ways.

Although the academic literature is replete with books on Jewish intellectual life, madness, and books on the state of the university, no book to date makes connections between these three topics as inter-related TROPES. Similarly, books on strangeness and Otherness abound. Maxine Greene's (1973) early work on *Teacher as stranger* is a precursor to this book. Greene's interests center on the notion of the "stranger" from an existential position. This book does not deal with the notion of "stranger" from an existential position, but from a psychoanalytic one. Moreover, unlike Greene, the notion of strangeness here is examined from the Jewish perspective. Svi Shapiro's (1999) book *Strangers in the land: Pedagogy, modernity, and Jewish identity*, is also a precursor to this book. Unlike this book, Shapiros's work does not look in depth at European Jewish intellectual traditions and does not tie Otherness to madness or the university.

Sander Gilman's (1982, 1993, 1994, 1995a, 1995b, 1997, 1998, 1999) work has been of tremendous import here. His interests in Jewish studies, madness, and the problems of intellectual history are groundbreaking. Unlike Gilman, however, this book's primary focus is education. Because of the focus on education, the insights suggested here—on the interrelations of Jewish intellectual history, madness, and the problems of the dystopic university—are unique.

Another book distantly related to this project, is Peter Trifonas's (2003) edited collection titled *Pedagogies of difference: Rethinking education for social change*. Trifonas argues for building coalitions across difference. Trifonas is mainly interested in discussing various kinds of differences and suggests that all kinds of people need to work to bridge gaps so that we can begin talking to one another across difference. This book does not focus on bridging gaps across difference, rather the project here is in exploring—in depth—Jewishness as a site of Otherness. Bill Readings (1996) *University in ruins*, is another book that is similar to this one. He argues that the university, because it has been fully corporatized, is in ruins. Readings is right on the mark. However, if we are to talk of "ruins" we must talk more psychoanalytically and explore what it *feels* like to experience the "ruins." Readings does not deal with this issue psychoanalytically or phenomenologically. Since the university is a dystopic place, a more psychological and phenomenological approach seems necessary. Moreover, Readings is not interested in dealing with issues of the university from a Jewish perspective whereas this one does.

Psychologically speaking, this work draws much from that of Michael Eigen's (1993, 1996, 1998, 2001a, 2001b, 2004). Of course, this book differs from Eigen's because the topic at hand is not only psychoanalytic. The focus here intersects psychoanalysis and curriculum studies. William F. Pinar's (1975/2000, 1994, 1995, 1998,

2000a, 2000b, 2001a, 2001b, 2004) work in the field of curriculum studies has been the main guiding light here. Pinar's early piece titled "Sanity, Madness and the School" (1975/2000), is the basis upon which this book was written. This book is a response to Pinar's (1975/2000) early essay. Pinar's general interests turn on race, gender, and the internationalization of curriculum studies. Unlike Pinar's work, this book, again, focuses mainly on Jewish studies, psychoanalysis, and intellectual history. Finally, this book differs from what has been written about both within the discipline of education and outside the field. No scholar to date argues in a psychoanalytic-poststructural fashion that the university is dystopic.

Pedagogy, Curriculum, and the Basics: Reconstructing Curriculum in the Public Sphere

This book centers on pedagogy, curriculum, and the basics in order to reconstruct curriculum in the public sphere. Teaching is what we do everyday inside the academy. Teaching makes up the nitty-gritty of our lives. And if our teaching is related to our work as scholars—as it should be—our students might learn not only subject matter and core competencies, but—more importantly—learn the crucial importance of intellectual exploration and creativity. Moreover, students and scholars alike should work from within, as William F. Pinar (2004) argues, in order to better address pressing social issues. The purpose of pedagogy is to teach people to teach themselves. To do edgy scholarly work, students must learn to educate themselves. Students must ultimately take flight, find their own "lines of flight" (Deleuze and Guattari, 1987, p. 32) after leaving the academy to become teachers. We teach pre-service teachers—to become teachers—in order that they can better teach themselves and their students. Becoming a teacher means becoming an explorer, an intellectual creator. Students might begin to think of learning as a large puzzle, putting pieces together and taking pieces apart. We synthesize and analyze. Students must pay homage to the intellectuals who paved the way for them by studying scholarly texts carefully. Students might begin to think through texts, to make these texts their own through studied analysis. This is no easy task. Many cannot analyze. Many cannot make texts their own. Many students cannot make texts their own because, in part, they have not done their own autobiographic work. Many students simply do not know who they are or what they think themselves to be. So I ask my students to write autobiographies. They resist. They tell me that they do not know where to start or what to write. I tell them to start at the beginning. They resist. They resist talking about themselves because schooling is never about the self. They are not used to talking about themselves and have little conceptual tools with which to do so. This is where pedagogy begins, with autobiographics. One must have some understanding of the self to analyze the texts of others. And one must read others' work to come to some understanding of the self. Thus, students might study together others' texts to better understand the larger sociocultural scene in which they are living. The more one finds out about the self, the more strange the self becomes! As William F. Pinar (1994) suggests,

> In a certain sense this work often feels like a voyage out, from the habitual, the customary, the taken-for-granted, toward the unfamiliar, the more spontaneous, the questionable. (p. 149)

The voyage out is actually a voyage in and down into the rumblings of the unconscious. The more closely one looks at the self, the more eccentric one becomes, primarily because of unconscious processes.

Michel Serres (2000) contends that "the voyage of children . . . is the naked meaning of the Greek word pedagogy" (p. 8). We are all children, in a sense. Doing intellectual work demands a childlike sensibility, a childlike curiosity. Teaching children reminds us of our own childhoods. But we must somehow move beyond naive reverie to get to the tough issues at hand.

Teaching is an art, not a science. Teaching is not about methods, it is about living. Teaching is a calling, not a job. Teaching is a life, not a career. Teaching is noble. Teachers impact our youth in profound ways. Teachers shape the future of the country. Teachers shape our children. Teachers inspire youth. Teachers are on a journey to know the self in order to better understand the Other. Pedagogy is an ethical calling. Pedagogues must be responsible to the young so that they may find their callings and live their lives the best way they know how.

To understand self and other, one must study curriculum historically, culturally, psychoanalytically, philosophically. Curriculum *is* lived experience. It is experience in the classroom and outside the classroom. Curriculum is subject matter across boundaries, across disciplines. Curriculum is a field of study that moves across fields, that examines educational experience generally speaking. William F. Pinar (2004) states,

> Curriculum conceived as currere [the Latin root of the word curriculum] requires not only the study of autobiography, history, and social theory, it requires as well the serious study of psychoanalytic theory. (p. 57)

Taking Pinar's lead, the issues in this book are explored psychoanalytically because psychoanalysis helps sort out complex emotional topics unlike any other theoretical framework. Throughout this book psychoanalytic insights are woven into the analysis to make sense of self-formation and Otherness. Peter Rudnytsky (2002) speaks to the importance of thinking psychoanalytically. He states,

> The appeal of psychoanalysis as a guide to living stems ultimately from the way that it enables its adepts to think theoretically about their own experiences. It thus functions, as it were, on a meta-level, not removing one from life, but immersing one in it more deeply. (p. 107)

Psychoanalysis is, as Rudnytsky states, a way of life; it is indeed a guide for living. Psychoanalysis is an interesting way to think through intellectual and emotional problems on a daily basis. Educational issues are also psychoanalytic.

Dwyane Huebner (2000a) remarks that "education is a manifestation of the historical process, meshing the unfolding biographic of the individual with the unfolding history of his society" (p. 246). The subjective and the collective move together in an unfolding of being and time. The time is now, but the time is also of the past and future. It is important to study issues historically, especially in the field of education since the field has been so plagued by ahistorical work. This book is historical therefore. The Jewish intellectual tradition is treated historically to better flesh out relations between past and present. This book is not about offering lessons to teachers about teaching Jewish studies better, or teaching Jewish children better, or teaching Christian children to tolerate their Jewish friends. These are all fruitless endeavors. Books that focus on teaching subject matters better or more efficiently are inherently anti-intellectual and ahistorical. Hans Georg Gadamer (1995) teaches there is little truth in method. If we teach method, if we teach in ahistorical and atheoretcial ways, we "infantilize our students" (Pinar, 2004, p. 157). William F. Pinar (2004) claims,

> The innocence with which some teach courses in teacher education not only infantilizes our students, it leaves them unprepared for the often contestatory character of public school classrooms. (p. 157)

Curriculum theorists try to complicate the conversation (Pinar, 1995), not dumb it down. Curriculum, the counterpart to pedagogy, is symbolic (Pinar, 2004). Curriculum symbolizes the myths and stories we wish to impart to the young. Curriculum is the psychic space whereby we try to figure out who we are against tradition. Curriculum is the Id, the ego and super-ego writ large. Curriculum is fantasy. Curriculum is a dream or a nightmare. Curriculum is the site of omissions and inclusions. Curriculum is to be theorized, not to be taken at face value or taken-for-granted. Curriculum is a living being; it changes over time and reflects what each generation wants it to reflect. Curriculum is a cultural mirror. We can look into the mirror and see both the beauty and the beast of the past and the present. Curriculum is the promise of our future. Curriculum is our legacy and our children's inheritance. Curriculum is, as Derrida (2002c) might put it, our "democracy-to-come" (pp. 204–205).

Curriculum is about the basics. David Jardine *et al.* (2003) remark,

> Imagine if we treated these things as the basics of teaching and learning: relation, ancestry, commitment, participation, interdependence, belonging, desire, conversation, memory, place, topography, tradition, inheritance, experience, identity, difference, renewal, generativity, intergenerationality, discipline. (p. xiii)

Following Jardine, this book (which deals with many of the issues Jardine raises) is about the basics. What could be more basic than trying to understand who we are? What could be more basic than trying to understand self and other? What could be more basic than studying Otherness and tradition? What could be more basic than studying issues historically and culturally? What could be more basic to education than examining the psyche and its discontents? Certainly these are the A B C's of curriculum theorizing.

Reconstructing Curriculum: Otherness and Public Education

Otherness tends to be a vague signifier. Otherness is trendy academic jargon. But in this book I do not want to be vague and I do not want to be trendy. When I talk of Otherness, I talk of Otherness as it is experienced concretely. Otherness is not only a philosophic symbol or an abstract concept. No, this is a lived, real feeling. This is the feeling of discontent. This is the feeling of eccentricity. This is the feeling of groundlessness and anxiety. Students and teachers alike feel these things. These are the feelings that go unnoticed in a classroom; these are the feelings that we do not address on a daily basis with our students. Yet, these feelings get expressed in students' acting out, sleeping, staring out of the window and not showing up for class. Faculty members act out in different ways. We might show our discontent by becoming increasingly anxious and depressed. Otherness is not trendy, it does not feel good. To be eccentric is to be out of place. To live in that space of Otherness is to be not-at-home, to be always already a floating signifier. Otherness is not a state of grace, it is a state of alienation. Bill Readings (1996) comments that "the pedagogic relationship, that is, compels an obligation to the existence of otherness" (p. 189). Otherness is the heart of pedagogy and curriculum. Otherness is not something we can afford to throw out, ignore, or obfuscate. Otherness may consist of one's own Oedipal drama, one's family romance. Otherness may be felt as a pressure, an expansion, darkness or light. Otherness is a felt experience.

The Other is not an abstraction, just as Otherness is not an abstract state. But to think the Other, one either has to identify with the Other or dis-identify with the Other. If one identifies, psychologically, with the Other, the Other may become, in a sense, symbiotic with the self. How to avoid symbiosis? How to avoid too close an identification with the Other? These are not easy psychological moves. One must constantly work at keeping the Other Other, keeping the Other at a distance so as not to collapse Other and Self, so as not to collapse subjectivities. Alterity is absolute, the Other is Other than self. And we must work intellectually and emotionally to keep things this way. Otherwise, we write only out of our own psychological projections. In one sense, there is no way to get around projections, but we can minimize these if we keep in mind that alterity is incompatible with symbiosis.

The tragedy about education—as a field—is that it has not always been possible to discuss Otherness. Education as a discipline has historically been reactionary. Otherness as a topic was not embraced fully until the 1970s with the advent of the Reconceptualization (see Pinar *et al.*, 1995). Likewise, public schools have always been a site of conformity and rigidity in America as well as in Eastern and Central Europe. Public schools are not places where children can be different or express difference or Otherness. Peter Trinfonas (2003) rightly points out that

> educational institutions have traditionally not tolerated the value of subjective differences among student populations. For the sake of securing the reproduction of "cultural capital" society and its normative ideals and models, the institution of education in the West has promoted the vision of a relatively homogeneous community. (p. 2)

Trifonas raises serious questions about what it is we think an educated community is or what it should be. The word community means to commune. But can we commune together across difference? Trifonas argues that we can, and this is what we should work toward as educators. But this idealistic vision is not so easy. Kids who are pushed around in school do not reason with their peers, they kill each other. Adults do not send a good message either. Just look at the war with Iraq. No talking, shoot first, ask questions later. Michael Moore's (2002) film *Bowling for Columbine*, certainly sends this message. Educators' work has only just begun indeed. And yet, scholars try to think through the notion of the Other and Otherness as a subjective experience. When students arrive in our classrooms, they are to be thought of as the totally Other. If teachers think they know their students they are mistaken. We do not know our young people. Just look at Columbine!! This was certainly unexpected. We must work hard to understand our youth; we must work to understand the Other and her subjective experience of Otherness. But this does not mean that after long study and intellectual exploration of culture that we understand clearly. We may understand bits and pieces of culture and history, but we are circumscribed by our psychological interests and capabilities. Moreover, Otherness is not merely a solipsistic state. Otherness is experienced relationally. A self is only a self in relation to the Other. Otherwise, there would be no self. This is the lesson that object-relations teaches (see, e.g., Rank, 1996; Klein, 1952/1993, 1957/1993, 1955a/1993, 1955b/1993).

When we are in-relation to each Other, we are squarely located in the public sphere. Private lives are lived in public spaces (Sumara, 1996; Atwell-Vasey, 1998; Pinar, 2004). When teachers and students read texts together, private interpretations of texts become public. What could be more public than teaching? What could be more public than public education? William F. Pinar (2004) puts it this way,

> If public education is the education of the public, then public education is, by definition, a political, psycho-social, fundamentally intellectual reconstruction of self and society, a process in which educators occupy public and private spaces in-between the academic discipline and the state (and problems) of mass culture, between intellectual development and social engagement, between erudition and everyday life. (p. 15)

To educate others is most fundamentally a public act. As William F. Pinar suggests, this is no easy task. The call to erudition, especially for public school teachers and professors of education is vital. We cannot afford to be anything less than erudite, especially in these post–9-11 times. The call to study politics and society, the call to study self-formation and Otherness has never been so urgent. The reconceptualized field of curriculum studies since the 1970s (see Pinar, 1995) has been a call for erudition. This erudition means that scholars responsibly educate young people so that they can lead us into the next century wisely and prudently.

Teaching pre-service teachers to want to become erudite is the task at hand. Traditionally teacher-education programs have not done this. Teaching in public schools, historically has not been viewed as an erudite profession. We must instill in our young people the call to public responsibility, to respond wisely. Teaching is a public act, and teachers affect children in complex ways. There is a grave ethicality facing us in the twenty-first century. This ethicality beckons us to teach knowledgeably and responsibly. Reading deeply and widely in one's discipline—reading deeply and widely in interdisciplinary ways—is reading ethically. It is simply irresponsible to do otherwise. Dwyane Huebner (2000b) states, "The struggle to remake the school is a struggle to make a more just public world" (p. 273). Curriculum professors struggle to make a more just public world by teaching pre-service teachers to become erudite. Complex concepts such as "justice" demand a studied response. It is not enough to simply spout rhetoric and slogans. We must deconstruct complex terms like justice with our students so that they understand fully their responsibility as public servants. The study of justice is always already one that is steeped in history and politics.

Who do we believe ourselves to be? That, of course, is a highly cultural question. The question is already loaded because we are thrown into a culture that tells us who we should be. It is against this "should," that erudition works. Erudition shows us ways out of the "should." Erudition gets us out from under the sedimentations that make up culture. Erudition can free us, somewhat, from who we think we "should" be. Erudition can make us better people, can free us from thinking that we can only be the way we are. Studying helps to unshackle the mind from the burden of cultural sedimentations. Studying helps to extricate one from the family romance writ large. The family romance writ large is the oppressive patriarchal drama of American culture.

Erudition can bring about new public spaces, public spaces that are somewhat more free. Jacques Derrida (2002c) calls upon us to "find the best access to a new public space transformed by new techniques of communication, information, archivization, and knowledge production"

(p. 203). New public spaces are those in which students and teachers can explore ideas, dreams, and public policies that might make for a better democracy. But this better democratic space in which we dream about will take study, work, and action. Most of all it will take a constant vigilance. Jacques Derrida and Bernard Stiegler (2002a) argue that "it is necessary, always if it is necessary—to try to train and to educate as many people as possible . . . to train them to be vigilant, to respond, and on occasion to fight" (p. 67). Professors and students alike must continually fight for academic freedom, to fight for the right to intellectually explore what moves them emotionally and personally. For many professors, though, this freedom is constrained. Academic freedom does not always mean total freedom. We are not free to teach as we wish, to teach what we wish totally. We must constantly fight to teach what we wish, what we deem necessary, what—in our best judgment—serves the public. To serve the public is our job. But often our service is shut down and "lines of flight" (Deleuze and Guattari, 1987, p. 32) are hampered by the university. What we must fight for is what Derrida calls a "university without condition." Derrida (2002c) argues,

> The university without condition does not, in fact, exist, as we know only too well. Nevertheless, in principle and in conformity with its declared vocation, its professed essence, it should remain an ultimate place of critical resistance—and more than critical—to all the powers of dogmatic and unjust appropriation. (pp. 204–205)

The key statement here is that the university without condition *does not exist.* Professors teach in institutions that are conditional. Scholarship is conditional. Who has the power to decide what knowledge counts? If a scholar's work is deemed too political or too radical it will not count in an institution that values tradition over against intellectual exploration. The search for knowledge should be an unconditional search. But it is not. The search for truth(s) is highly political. These problems are not new by any means. But they continue to plague the university. If the university serves the public by teaching tradition, how can it "make room" (Eigen, 2001a, p. 5) for the new, for intellectual exploration? Does the university "make room" for new knowledges? It must, but it does not always do so.

If we are to reconstruct the public sphere of education we must "make room" (Eigen, 2001a, p. 5) for new knowledges *unconditionally.* If we are to reconstruct public schools, we must allow teachers to teach what knowledge they deem important; we must empower our school teachers to become erudite, to take back the curriculum, to allow our children to flourish, rather than perish in the rubble that is Columbine. Teaching teachers takes on a new urgency when talking of reconstructing a new public sphere of education. Adam Phillips (2002) contends,

> The idea is not to sit around all day having interesting conversations, and entertaining points of view. It is with a sense of some kind of urgency—with a sense of something being at stake, of something that can't be set aside or ignored—that people enter the political arenas, [I would add educational arenas] and consult their respected professions. (p. 15)

Education is not entertainment, though many of our young students would like nothing more than to be entertained. Education is tough, it is hard work, it is the task of thinking through the sociopolitical as well as subjective experiences. Reconstructing a new public space of education is an urgent call. We simply must make changes in the ways we educate the young. Nobody was ready for 9-11. We are not ready for the future-to-come. We live in horrible times, we live in violent times, we live in a time of urgency. We must "make room" for this urgency by study. We still do not know who we are as Americans. We still have a lot of historical work to do to get some kind of understanding of why we are so hated abroad. We are a young country and our children's lives are at stake. Our very future as a democracy is in the hands of our children. But we must take our children's futures seriously. Roger Simon (2000) asks,

> What might it mean to live our lives as if the lives of others truly mattered? One aspect of such a prospect would be our ability to take the stories of others seriously enough, not only as evocations of responsibility but as well as matters of "counsel." (p. 62)

Reconstructing the public sphere of education means taking the lives of others seriously by seriously listening to them. Part of the problem with public education is that the larger public is not listening to the pleas of our schoolteachers and our children. This willed ignorance negatively affects all of us. What will it take for the larger public to listen? Another Columbine? We do not need more standardized testing. Standardized testing is a form of violence against children. Standardized testing is the fear factor in children's lives. Education is not about standardized knowing, it is about erudition and taking the lives of others seriously. Reconstructing the public sphere in education means opening up new public spaces for all of us. The call for the unconditional university is one step. We must also call for the unconditional public school. Public education at all levels must be unconditional.

The intellectual journey in this book is meant to open up the conversation on reconstructing the public sphere in education. The book begins by taking seriously the words of Jewish intellectual ancestors and taking counsel from them, as Roger Simon suggests. Part 2 examines psychic upheavals reported by those who have experienced extreme states of Otherness and part 3 takes seriously thinking through the dystopic university. This book is an intellectual journey into the land of Otherness. The three tropes of Otherness (that of the Jewish intellectual, that of madness, and that of the university) might open conversations on what it means to write psychologically, phenomenologically, and poststructurally.

CHAPTER 2

Jewish Intellectuals as a Trope of Otherness

The aim of this chapter is to examine Jewish intellectuals as a trope of Otherness. Jewish intellectuals have always lived on the margins, have always experienced alterity. Jewish intellectuals are outsiders to mainstream culture because (1) they are Jews and (2) they are intellectuals. Historically, Jewish intellectuals have expressed a feeling of Otherness because of anti-Semitism. In this chapter, European Jewry and its relation to American Jewry will be examined—especially as a trope of Otherness.

Phantasies, Public Spaces, and Scholarship

Let us begin this chapter in the company of Jean Laplanche (1999): "let there be an element of this work worthy of being archived, and it—once in the archive—will not be covered over forever with dust of never-borrowed manuscripts" (p. 8). Once a work is archived in the public space of publication it is of course open to public scrutiny and the public gaze, and for public university scholars and public school teachers alike to ponder. It is hoped that the work will be studied by those who dwell in public institutional spaces. Studying this text might allow scholars and teachers to create better public and private spaces. Better spaces are created ethically. Ethicality always entails a conscious sense of responsibility of the call-of-the-Other. The private and the public are not two separate realms but bleed over into each other. The private lives of teachers and scholars bleed over into the public life of teaching inside public institutions. Drawing on the groundbreaking work of William F. Pinar (2001b), it is suggested here that studies like this one—of cultural, psychosocial, pedagogical, and curricular importance—might open public spaces to our colleagues so that they may form their own complex ideas about teacherly and scholarly lives. Pinar (2001b) eloquently states thus:

> Employing research completed in other disciplines as well as our own, let us construct textbooks . . . which enable public school teachers to reoccupy a vacated "public" domain, not simply as "consumers" of knowledge, but as active participants in conversations they themselves will lead. In drawing—promiscuously but critically—from various academic disciplines and popular culture, I work to create a conceptual montage for the teacher who understands that positionality as aspiring to create a "public" space. (pp. 21–22)

This publication creates open intellectual terrain that moves between the private and public, between tensions created by an oppositional stance that one must take against those who cannot embrace difference. One must create an open intellectual space, an open intellectual terrain in order to do the work. Hannah Arendt (1958/1998) remarks that "[t]he space of appearance comes into being wherever men [*sic*] are together in the manner of speech and action, and therefore predates and precedes all formal constitutions of the public realm" (p. 199). How one "appears" to another in the public realm is highly complex both sociologically and psychologically. The ways in which one "appears" to another marks that relationality via difference. That is, the other is only known through her difference. And it is through that difference that public spaces and open terrain of intellectual work can be done responsibly.

Part of the complexity of relationality and intellectual work concerns the unconscious. This Otherness within (the unconscious) makes one strange to oneself and stranger to another. Jean Laplanche (1999) suggests that "the unconscious is not a corpse, it is a revenant" (p. 19). There are ghosts within the self and within the Other. We are haunted by our primary figures who get installed in our psyches at a very early age. The ghosts of primary others get transported from one realm to another—from other to self and from self to other—by complex mechanisms of incorporation and projective identification respectively. The ghosts of the unconscious hover and haunt. Naomi Rucker (1998) comments that it was Ferenczi who first pointed out that the unconscious is "cocreated." She says, "The idea that the unconscious is shared, is shareable, and is mutually constructed or cocreated in both the original developmental relationship and in the psychoanalytic field is traceable directly to Ferenczi (1915)" (p. 22). If the unconscious slips between infant and mother via introjective and projective processes, it can operate similarly through infant and father, adult and adult, teacher and student. The psyche, thus, is formed in a complex matrix that moves uncannily inbetween introjective and projective spaces. Taking things in psychologically and spitting things out are fundamental to the human condition.

Naomi Rucker and Karen Lombardi (1998) make an interesting claim about the dynamic nature of the unconscious.

> We wish to move beyond the concept of the unconscious merely as that which is dynamically repressed or as that which holds dissociated contents, to a concept of the unconscious existing in dynamic relation to conscious processes, serving a linking or translating function between the internal and external worlds, between self and other. (p. 10)

This linking or translating function of the unconscious seems to be something about which scholars know little. The lack of linking may be the root of schizophrenia or other psychotic illnesses. If one cannot connect or link to the outer world, one remains stuck in a state of solipsism and perhaps even autism.

When links are made, though, translations of those links are messy and unclear because the unconscious is messy and unclear. It is the unconscious which drives us to do what we do and write what we write and what we create, yet there is no getting a handle on why it is we are driven the way we are. Of course, there are clues along the way. On a conscious level, one knows why one is interested in certain scholarly topics, but on a deeper level it is difficult to say why we write on what we do. Georg Groddeck (1961) argues that "one cannot speak about the unconscious, one can only stammer" (p. 26). Groddeck influenced Freud's ideas about the Id. Unlike Freud, Groddeck called the unconscious the "It." Groddeck (1961) remarks,

> I hold the view that man is animated by the unknown, that there is within him an "Es," an "It," some wondrous force which directs both what he himself does, and what happens to him. The affirmation "I live" is only conditionally correct, it expresses only a small and superficial part of the fundamental principle, Man is lived by the It. (p. 11)

The "It" is not a thing, it is not a place, it is not Freud's Id. It is something more mysterious, more mystical, more mythical, more interesting. Lawrence Durrell (1961) asks, "And what of the It? Groddeck [insists] . . . that the It is not a thing-in-itself, but merely a way of seeing" (xiii). "It" is a way of seeing, "It" is way of being. The

"It"—as Groddeck conceptualized it—is messy. Groddeck (1961) says "For the It there exists no watertight ideas, it deals with whole structures of ideas, with complexes, which are formed under the influence of symbolization and association" (p. 48). Processes of symbolization and association create another layer of thought that complexifies lived experience. Symbolization and association are dense. The other *symbolizes* someone—perhaps a parent figure—to the self. When one meets a stranger one might be reminded, say, of one's father. The *association* of the other with one's father might trap one into old patterns of relation with the father. The other as symbol to the self is also a symbol of the self or of parts of a split-off self. The self is always already a symbol to the other.

One's research and scholarship follows similar trajectories. That is why it is difficult to say why certain topics are of interest and others are not. One reads certain texts because they remind one of a primary figure. One reads certain texts for psychological reasons not completely understood; one avoids other texts for psychological reasons as well. Texts take on symbolic meaning just as people do. But these symbolic meanings are nearly incommunicable. Attempts to communicate that which is incommunicable become highly problematic as the internal psychic archive is a tangled weave. Laplanche (1999) explains that the unconscious has nothing to communicate, rather it is through the process of analysis that the unconscious speaks, and yet it has no message. Laplanche explains this conundrum:

> It might be said—perhaps metaphorically—that the unconscious speaks, but it does not aim to communicate, it does not convey any message. In the dream, in the symptom, even in symptoms which are directly linguistic, such as slips of the tongue, we have to insist on the following. They are formations of the unconscious, closed in upon themselves, and not intended for communication. (p. 103)

Like analysis, scholarly writing—which is a form of textual *analysis*—is *meant* to communicate messages that get tangled up in our unconscious. But it is a puzzlement whether the text communicates anything. Is anything ever communicated? Lacan teaches that we never understand one another, that there is no such thing as transparent communication. But that does not mean that scholars do not speak at all. Whether the message-text is understood or not is always already a question. When teasing out patterns across texts via free associative processes, scholars speak/write (both are tangled up together) through complications. Some things might be understood by others, and some things might not. Some ideas get blurred, while other ideas make sense. Much of what gets understood and worked on depends on who is reading the text of course. Scholars cannot expect their work to ever be fully understood. Misunderstandings are to be expected because both the writer and the reader operate out of the unconscious. The unconscious has a mind of its own—as it were.

Scholars work in the realm of phantasy. That is, they take in and spit out thoughts that are driven by the unconscious. Because scholarship is mostly phantasy, this fact doesn't make it any less real—indeed it is real. But the process of doing scholarly work is not fully understood. Phantasies are grounded in unconscious slippage. This slippage surfaces in the form of dreams as well as in intellectual work. Melanie Klein (1946/1993) argued for the primacy of phantasy in the psychic life of both children and adults. Slavoj Zizek (1997) says "fantasy tells me what I am to my others" (p. 9). Phantasy life complicates the web of all of our relations. Phantasies shape writing and living, being and doing. The phantasy of who the other is is perhaps more real than who the other *really* is. But who the other really is cannot be gotten at if phantasy is the ground of experience. If phantasies make up large parts of our psychical life, then attending to psychical life is important. Freud (1935/1989) pointed out in his autobiography that "psychical reality was of more importance than material reality" (p. 37). What Freud is not saying is that psychical reality is not real because it is psychical. Material and psychical reality are both real, but both operate on slightly different registers. And these two registers slip and slide over against one another, on top of one another and between one another. Psychical reality colors material reality; material reality colors psychical reality. Groddeck (1961) suggests more radically the following: "But whoever is familiar with the kingdom of phantasy recognizes, at one time or another, that all science is a kind of phantasy" (p. 8). Donna Haraway (1989) in her book on primate research has pointed this out. Feminists especially can attest to the phantasy that is science. Again, this does not make science or social science, for that matter, less real. Phantasy, symbolization and association, projective identification, and introjection drive scholarship. It is crucial to understand that much of psychical reality is unconscious. In fact, Julia Segal (1995) points out that "phantasises are by definition unconscious fantasies: you cannot simply look into your conscious memory and say" (p. 36). One only has vague hints and clues about what phantasy life means. Julia Segal (1995) remarks that "We see people [or texts written by people I might add] through phantasies which are not only the end-result of our learning but also a lot of working out in our heads and a lot of distortion of and addition to perceptions of internal and external reality" (p. 28). Internal and external reality, though, are not two separate, clear realms. Both are intertwined and confused with one another. Karen Lombardi (1998) points out that "[n]either phantasy nor external reality is absolute. External reality is not just objective and psychic reality is not just subjective; rather, both are versions of reality" (p. 101). The important statement here is that psychical reality is a *version* of reality. Commonsense understanding of phantasy is that it is all make believe. Well, to a certain extent it is, but at a deeper level it may not be. Phantasy may be based on very real experience, or phantasy may be based on something surreal, more than real. Or it could be based on nothing. Whatever phantasy is, it is based on a person's perception, whether or not it is real to another person. As Hannah Segel (2000) points out it was Melanie Klein, not Freud, who argued that phantasy emerges as a psychological constitution at the very beginning of life. Freud, unlike Klein, believed phantasy happens later in life when thought becomes organized and structured. Segel (2000) teaches that

> it is the psychoanalysis of children which revealed the ubiquity and the dynamic power of unconscious phantasy, and Klein gave this concept its full weight. From the beginning of her work with children, she was struck by the extent to which the child's life is dominated by unconscious phantasy. (pp. 18–19)

Not only is the child's life dominated by phantasy, adult's lives are also dominated by phantasy, albeit of a different sort. Perhaps adult phantasies are more nuanced. Our parents, siblings, peers, and even the institutions in which we are housed become part of phantasy life. Not only that, phantasy guides our intellectual labor. The world one encounters through lived experience and through the study of texts is a world that appears through the lens of phantasy. This lens is murky and ghostly. Freud argued that phantasy served several purposes simultaneously. Mainly for Freud, phantasy served defensive functions against painful memories. Why the defensive function? Phantasy occurs when reality pressures the self to retreat. Psychic retreat, though, does not always mean becoming pathological or becoming ill. Psychic retreat into phantasy may be a daily occurrence. Schools and universities by their very nature push people into retreating into phantasy life because of their deadening atmosphere. Life within these institutions foster phantasy by their very nature because academic institutions are not human. Humans long for companionship—intellectual companionship. Schools are certainly not the places where children find intellectual companionship. Likewise, in university settings, intellectual companionship may be lacking. This is not a new phenomena by any means. Jewish intellectuals like Walter Benjamin and Hannah Arendt shuddered at what went on in European universities. Walter Benjamin (1994), in a letter to Ernst Schoen, states,

> The only salient point—which you know more deeply because you have never experienced it the way I have—is that this university is capable of poisoning our turn to the spirit. On the other hand, this is the only salient point: that I made the decision to run the gauntlet of the course of lectures . . . and saw the shrill of brutality with which scholars display themselves before hundreds of people; how they do not shy away from each other, but envy each other; and how ultimately, they ingeniously and pedantically corrupt the self-respect of those who are in the process of becoming. (p. 74)

The unprofessionality of professors—who act like devious children—as Benjamin comments upon, is rather shocking especially for people on the outside looking in. Seasoned professors might not be shocked by all of this, but certainly junior professors might be taken aback. This kind of atmosphere prevents one from finding intellectual friends. Ironically, Benjamin (1994) states, "[t]he university is not the place to study" (p. 72). If the university is not the place where one can study, where does one turn? Benjamin, of course, turned to cafe life in Paris and other interesting sites like arcades, tattoo parlors, and circuses to study popular culture. Benjamin's university was the street. Benjamin became an important cultural critic and one of the most interesting cultural studies writers of the twentieth century.

But today being an independent scholar—like Benjamin—who is not housed in a university is simply not financially possible, unless one is wealthy. One simply cannot make a living from scholarly treatises; scholars need the university for financial support. So with or without intellectual companionship—inside the university—scholars push onward. The university can be a lonely place for scholars longing for intellectual friendships. When one is lonely, phantasy ensues. When intellectual companionship is lacking, phantasy fills the void. Hannah Segel (2000) contends that "Phantasies of course are linked with defenses. The very fact of phantasying is a defense against painful realities" (p. 21). In the space of university life, scholars phantasize about other intellectuals to whom they can speak. Thus, scholars must read books in order to find intellectual companions.

Sophie Freud (2004), in a marvelous piece titled "The Reading Cure: Books as Lifetime Companions," remarks that she

> reads books for companionship, for enjoyment, for solace, for information, for distraction, for self-improvement, for self-knowledge, for understanding, for enlarging my world, for enhancing my compassion and empathy with totally different others. (p. 77)

Sophie Freud talks of books as "companions." Indeed. Interestingly she tells us that she engages in what is called "bibliotherapy—healing through reading books" (p. 83). Reading to heal, to find friends, to find companionship may be the reason why many do intellectual work to begin with. Ironically, the university is a desert. It is not a healing place or a place where one might find solace. There is little peace to be found there. There is little intellectual stimulation in the everyday life of the university professor. Intellectual life at the university is squashed by the demands of service, by the demands of the trivial.

Otherness and Jewish Intellectual Ancestors

A study—such as this one—of the Habsburg Empire and beyond, might help scholars to better understand the troubling problems that Jewish intellectuals faced in Europe. Jewish intellectuals might then begin to understand how deep and how long the history of hatred, of anti-Semitism has been. This inquiry is broad and not by any means all encompassing. This is a sketch, a portrait, a general picture of European intellectual life during a broad timespan. I will not be dealing with the Holocaust in this book because I have already treated this subject in depth elsewhere (Morris, 2001). The focus here will turn on the years that preceded the Holocaust. Some questions that will be raised are these: How did Jewish intellectuals survive European anti-Semitism and keep writing? How did they avoid becoming cynical or nihilistic?

I state here that I am not a historian. Thus, I do not follow a strict methodology that a historian might. My field is curriculum studies. So I approach my study from a curricular perspective. Therefore, questions around the curriculum writ large are of primary import. What is of interest are the broad relations between lived experience, culture, school and university life. Of interest here is exploring, generally speaking, the lives of intellectuals. Hence, this section of the book presents a montage of Jewish intellectuals who lived in Central and

Eastern Europe during the years that led up to the Holocaust. A question that is explored in this book is how educative institutions shape Jews, how politics and the larger sociocultural arena impacted Jewish thinking and writing. Again, the main question turns on this: How is intellectual work possible against the backdrop of hate, against the backdrop of anti-Semitism?

The first section of this chapter will look at Jewish identity. The second section of the chapter will serve to contextualize historically and socially what it meant to be Jewish within the larger sociocultural arena of Eastern and Central Europe during the years that led up to the Holocaust.

One of the areas that will be treated turns on Austrian culture and politics. It must be mentioned at the outset that Austria serves as a special case when talking about European anti-Semitism, for it has always been one of the worst hotbeds of anti-Semitism in Europe and has had one of the worst records of anti-Semitism in all of Europe. Surprisingly, many argue that Austria's record of anti-Semitism is worse than that of Germany's. Amos Elon (2002) claims,

> The status of Jews in Austria-Hungary was also more worrisome than in Germany. The Austrian Social Democrat Karl Kautsky wrote that Austrian anti-Semitism was more dangerous than the German variant because of its pseudo democratic cast, which appealed to workers and oppressed nationalities. Visiting Austria in 1910, the scholar Victor Klemperer claimed that whereas he felt "abroad" in Italy or France, in Austrian ruled Bohemia he sensed he was in "enemy territory." Anti-Semitism, unequaled elsewhere in the West, was said to be rampant there. (p. 252)

One has to remember that Hitler came from Austria and learned what he did from other Austrian politicians. As I suggest in my work on the Holocaust (Morris, 2001), many historians point out that Austria had a disproportionately large number of Nazi hit men [and women for that matter] as compared to Germany and other European nations. As well as the trouble that is Austria, I have discovered—through my work here—that Jews, living in Romania, Poland, Czechoslovakia and, of course, Germany (not to mention all the other European nations that became collaborators with the Nazis during the Holocaust), had to contend with a great deal of anti-Semitism *in schools as well as in the universities.* One might think in places where learning goes on, that anti-Semitism might be frowned upon. But just the opposite is true. It seems that for many of these nations, schools and universities were also hotbeds of anti-Semitism. In fact, much of

Gentile youth bullied Jews, while school teachers and professors either turned their heads or—shockingly—more often than not—joined in.

Curriculum and Jewish Identity

Before moving directly into the section on Jewish intellectuals in Europe, it is important to raise some curricular questions that intersect with identity issues. Curriculum concerns the complex intersections between students, teachers, and texts. Students who seek to discover more about their identity-formation may do so via textual studies. Do we not read books in order to discover who we are? As Dennis Sumara and Brent Davis (1999) point out, "the search for knowledge is a search for identity" (p. 55). The search for knowledge may also be a search for dis-identity, or dis-identification. Dis-identification is akin to difference. When one dis-identifies with what one is reading one becomes uncomfortable. Difference makes one uncomfortable. Students and scholars may use defense mechanisms to block the incorporation of difference into their psyches, or they may shut down altogether and stop reading. Contrarily, students and scholars alike identify with certain texts. To build up the ego, one may read books with which one can identify, or one may lose oneself while reading a text. To dissolve the ego while reading is the feeling of a loss of boundaries. Reading texts and doing textual studies impact self-formation because the text does something to the reader as the reader reads the text. The reader moves in psychic waves between a "no" and a "yes," a taking in and a spitting out, an incorporation or a refusal. Maybe the issue is not even what the text does to the reader or what the reader does to the text but that "something" inbetween reader and text happens during the process of sense making. What this "something" of sense making is, is not clear. When the purpose of reading is self-formation or the discovery of difference, the process is intimate and profoundly phenomenological. Relationships between reader and text are not unlike those between two people. I take my cue from Naomi Rucker (1998) as she interestingly remarks,

> The recognition that our intimate relationships are shaped by the figure and ground of our known and unknown selves and the known and unknown selves of the other, rather than by interactions between wholly consolidated individuals, positions us to experience relational shifts from a creative vantage point. Our intimate experiences [I would add our intimate experience of doing textual studies] are no longer confined by the understandings that we are doing something, that something is being done to us, or even that we are doing something

> together to each other. Rather, it begins to reveal the possibility that we are doing something *through* each other and that something is being done *through* us. (p. 128)

The unconscious part of the reader is the unknown part. The unconscious part of the writer gets somehow injected into the text. It seems that these two ambiguous psychic relations (of reader and writer) get intertwined. The movement of self-formation involves psychic slippage through, between, and across texts as a vacillation occurs between finding, losing, projecting, and introjecting that which one is more than likely unaware. Readers take in what they can take in, spit out what they must. Sometimes we read poorly, not being able to pay attention to our conscious thoughts, daydreaming while reading. Yet, some of the daydreams that conceal thoughts slip through to the unconscious and appear later in split-off forms. It is later, in the dreamwork, in working on the meaning of the dream, that one might gain insight into associated meanings of reading.

Reading, the search for knowledge and understanding, and the love of books is particularly relevant to Jewish culture. In fact, Judaism is often considered a religion of the book. Jewish scholars, therefore, are able to make an easy transition from religious reading practices to scholarly ones. Susan Gubar (1996) explains thus:

> Clearly sharing a devotion to the book, which may have been fostered by a religion based on reading, interpreting, blessing, kissing, and parading classical Jewish texts, as well as an absorption with what Steiner calls "the unhousing" of language, many have extrapolated a career in letters . . . the orientation toward education in Jewish culture also brought many Jews into the academy. (p. 27)

Interestingly enough, however, studying one's Jewish identity as a scholar has not always been acceptable within the halls of academe. Today, however, it has become more acceptable to research Jewish identity-formation. Gary Saul Morson (1996) tells us that many Jewish American scholars are beginning to be more open to thinking about who they are and how self-knowledge affects their intellectual work.

> In ways that are often quite unexpected, many Jewish American scholars have found themselves listening to a Jewish voice within them that they had long neglected. The result is often a resurgence of some sort of Jewish consciousness, a dialogue of their ethnic or religious heritage with their professional training, which may be quite different from

> Jewish concerns. Like all true dialogues, these may yield insights that neither voice would have reached on its own. Identity acquires a new meaning and tonality, while professional communities may achieve greater depth and moral force. (p. 78)

Studying texts that directly impact one's Jewish identity does indeed deepen knowledge and understanding about the ways in which Jews historically have been othered and scapegoated. Studying texts that directly impact one's Jewish identity allows one to understand the complex mechanisms by which Jews must psychically manage tensions in a hostile environment. It is no great comfort, though, to know that being Jewish has never been easy and that prejudice and stereotyping are historically entrenched. However, it does, at least on one level, allow one to begin to understand how deeply historical these problems are and how prejudice and discrimination do not just go away with changes of attitudes or better economic times. The quest for understanding and becoming learned for Jewish scholars continues despite ongoing hatred. When a people are hated as much as the Jews, the quest for self-knowledge becomes more urgent; the search for identity becomes more intense. Thinking autobiographically, as William Pinar (2004) has pointed out, is key to doing good scholarship. For many Jews thinking autobiographically might mean thinking through what it means to be Jewish in a hostile world. Norman Finkelstein (1996) speaks to these issues eloquently.

> How much more loudly that voice speaks out when I attempt to write not criticism but autobiography, to write my life, that text with which I am most intimate. . . . From the Pirke Avot: "Turn it and turn it again, for everything is in it." From Edmond Jabes: "Thus the Jew bends over the book, knowing in advance that the book always remains to be discovered in its words and in its silences." So I study my life, as a series of tropes, which, taken together, make up the text of myself. (p. 415)

Jewish intellectual life, in this book, is also studied as a series of tropes. The major tropes of this book are tropes of Otherness. The major themes of this book are (1) Jewish intellectuals as a trope of Otherness, (2) madness as a trope of Otherness, and (3) the university as a trope of Otherness. These three tropes may be approached as separate issues, of course. But my aim is to argue that all three are interrelated.

In the field of curriculum studies, there is little material on Jewish identity and Jews are not often considered marginalized within the academy. Svi Shapiro (1999) points this out.

> It is also curious, if little commented on, fact, that while the "postmodern moment" has meant an unparalleled acknowledgment of the salience of "difference" and the "other" in the constitution of our social world, within critical educational studies at least, this has not included much about Jews as either an oppressed or marginalized group. (p. 31)

I hope that my contribution to the field of curriculum studies begins a conversation about Jewish intellectuals as marginalized and oppressed by the largely Christian culture in which Americans live. Jewish intellectuals are marginalized in the academy just as they are marginalized in the larger American culture. Thus, Jews are put on the defensive—having to figure out ways to psychically manage being othered. There is nothing new about this. David Bleich (1999) explains, "Having internalized a well-documented hostility from host societies, the oppositional aspect of being a Jew is familiar to me" (p. 111). Alan Block (1999) contends that not only is the Jew marginalized, he/she is absent from the curriculum altogether. Block states: "I would like to suggest that the Jewish voice not only has been silenced from curriculum studies . . . but also is absent from the curriculum in the school" (p. 163). Curriculum studies is a discipline that welcomes the other and encourages the other to speak, encourages the other to find a voice in the wilderness. And yet, it is surprising that many Jewish scholars in my own field—curriculum studies—do not address their own Jewish subject-positions.

Jewish subject-positions, of course, are not monolithic. Identity is complex. Perhaps, though, the word "identity" misleads. To "have" an identity suggests having something with a core or an essence. This is not my meaning. Identity is a psychical process, and a social formation. It is sociocultural and ever-changing. Marie Coleman Nelson (1997) argues that we are made up of multiple selves and that these are both intrapsychically and socially formed.

> I think of these selves as a harp, which is a vertical concept somewhat like Lacan's. Some things, some experiences which may be intrapsychically generated, or generated from the outside, strike certain chords, themes, or stances on this harp, and that is what you hear. (p. 87)

The self (s) speak in many registers. The key to understanding self-formation is understanding that the self is formed in relation to the Other, whether that other is another person, an introjected memory, or a text. Within and across these registers, Otherness pervades. However, one cannot fully understand who the other is, or what the author of a text *intends* to tell us. It is a very different thing to understand a text to the best of one's abilities and to get at the intention of what the writer means to say. This is because textual analysis is highly complicated and highly unconsciously driven. Otherness permeates textual registers. When we get down to it, it becomes difficult to say *exactly* what Otherness means. The best we can do is get at a vague understanding of what Otherness is about. Stefan Jonsson (2000) explains thus:

> As soon as we think we believe that we can think otherness, we can be sure that we are wrong, because the very act of thinking otherness implicates an assimilation of "the other" into "the same," into our own conceptual paradigms and ideological fantasies." (p. 141)

One's phantasies about the other and the other's Otherness perhaps tell us more about the self than the other. It becomes important to examine one's phantasies. As scholars, we try to do close readings to honor the text and the writer, but as close as we would like to read the text, we still read through our phantasy lenses.

Another complication of understanding self-formation concerns hyphenated identities. Jewish Americans, for example, straddle between two worlds. The tension between being Jewish and being an American varies depending on what it is that people tend to emphasize in their everyday life. Reformed, conservative, and orthodox Jews tend to emphasize one aspect of themselves over against another. Some Jews tend to find a balance between their religious beliefs and being American. David Gerber (1996) comments on his hyphenated identity in terms of doubleness.

> I felt within me two people simultaneously. I lived at once in two cultures that refused to be easily reconciled, and according to two codes, one of the public, cosmopolitan, and hopeful, and the other, private, ethnic, and nourished by a sense of being an outsider. . . . I craved this tension which I dimly recognized as a source of energy, imagination, and even wisdom in personal relations, conversation, and intellectual discourse. (p. 125)

When one lives in a culture that shuns difference and is hostile to Otherness, one does develop a double life. Part of the psyche goes underground in order to survive; part of the psyche must perpetually be split off in order to be protected from hostility. This is probably why Jewish jokes may come in handy psychologically, as Freud

(1960a) knew. Jokes can offset tension Jews feel from the outside world. Jokes are more than a laughing matter, however. As Freud points out, jokes spring from a deep unconscious site and operate like dreams. Freud (1960a) comments,

> To the totality of these transforming processes I gave the name of the "dream-work"; and I have described as a part of this dream-work a process of condensation which shows the greatest similarity to the one found in the technique of jokes—which, like it, leads to abbreviation, and creates substitute-formations of the same character. (p. 30)

The purpose of jokes, like dreams, is partially to protect the psyche from overloading. One jokes to get rid of psychic tension. One dreams what one cannot think. One jokes about what one cannot assimilate into consciousness. It is in the dream-work or the telling of the joke that tensions get worked through. Without jokes, Jews would be miserable. Jewish jokes are a direct response to anti-Semitism and hostility. So too is the work of scholarship and teaching for Jews who are willing to fight anti-Semitism actively.

Some scholars draw a distinction between ethnic and religious Jews. Sander Gilman's (1998) stance reflects my own.

> I write as an American Jewish academic, whose interests span everything from the history of medicine to the sociology of popular culture. My political and academic consciousness in the 1990s is rooted in my political education in the 1960s. Back then "Jewishness" as an ethnicity not only didn't exist but also was quite unacceptable. The American response to the Shoah had limited Jewish identity to religious identity. It was fine to be a "religious" Jew, but a secular "ethnic"—no, that was not possible. (p. 13)

At one level, it is surprising that it would be problematic to call oneself an ethnic Jew or to do scholarship as an ethnic, but not religious, Jew. But against the backdrop of the Holocaust, perhaps this position becomes a little more understandable. Many writers of Holocaust historiography and literature found that they had to deal with similar problems, especially during the early years of writing on the Holocaust. During the 1960s especially—when Holocaust scholarship was just getting off the ground—there were many taboos in the Jewish community about writing and doing scholarly work around one's Jewish identity and the impact that the Holocaust had on one's own identify-formation. Jewish victims and survivors' narratives were not found in historians' accounts of the Holocaust because it was thought that their testimony was unreliable (Morris, 2001). Some even thought it un-sacrosanct to use survivors' voices in their historiography. Likewise, some conservative Jews found it highly problematic to artistically represent the Holocaust, especially in the form of Holocaust fiction. As is well known, Adorno (1995) thought it anathema to write poetry about Auschwitz. Of course, these extreme—and I might add reactionary—attitudes have changed over the years. It also used to be the case, especially during the 1960s, 1970s and even well into the 1980s that American Jews could not criticize Israeli politics. Today, of course, the situation is different. Many American Jews are highly critical of Israel's human rights abuses of Palestinians. Of course, many Israelis, too, are disturbed and opposed to their governments' human abuses.

American Jews have complicated identities because those of us who are not directly related to the Holocaust, Gilman (1998) contends, lead some Israelis to argue that we are not really Jews at all. And if we are not really Jews at all, how can we write about things Jewish? How could American Jews, for example, write about the Holocaust? In this context, Gilman speaks out about the Daniel Goldhagen controversy (see for example, F. Little, 1997). Gilman argues that the controversy over Goldhagen's (1997) book *Hitler's willing executioners*, had little to do with the content of the book. Gilman contends that Goldhagen got severely criticized—if truth be told—because he was an *American* writing about the Holocaust. Many Europeans feel that Americans have no business writing about things European, especially the Holocaust. Goldhagen's (1997) book, *Hitler's willing executioners* was roundly criticized for errors, simplicity, and its major thesis. The thesis of the book is that Germans killed Jews because *they wanted to*, that Germans embraced what Goldhagen calls "eliminationist anti-semitism" (p. 132). Some critics read this as a gross essentialization. But Gilman claims that the real controversy has to do more with Goldhagen being an American Jew; it is his Americanness that infuriated critics, not the errors or essentializations in his book. Gilman (1998) forcefully states,

> In the reception of Goldhagen's book it is not the "Jews" in general who bear the brunt of these attacks, but "American Jews," defined in such a way as to define the absolute location of corruption and evil. (p. 189)

Gilman is one of the few scholars who has come out in support of Goldhagen's project. Many scholars got caught up in the Goldhagen scandal without really considering the larger cultural problem.

Although Goldhagen's book may be problematic, I do not think his work should be dismissed out of hand. (Morris, 2001). There is value in what he says, even if it is in need of qualification and nuance. What is more important here, though, is Gilman's argument about Goldhagen's Americanness as the backdrop of the criticism. Gilman makes a stunning claim indeed. Can an American Jew write successfully about topics such as the Holocaust without getting unfairly criticized, especially by European historians of German extract? I do not believe that anyone owns academic terrain. Americans can and should write about the Holocaust and should not be criticized for doing so just because they are American. It is crucial for Americans to think through the Holocaust for many reasons of course. It is terribly important for historians to write through their own cultural lenses to reinterpret the Holocaust for generations-to-come. The Holocaust is not the property of Europeans, it is not the property of anyone. The Holocaust is a burden for us all to think about.

The problematic nature of doing scholarly work as an American Jew, the problematic nature of being a minority and being constantly on guard against anti-Semitic slurs, creates ways of thinking about the world that are different from that of mainstream culture. When scholars are not steeped in the status quo and the mainstream, writing takes on a different tone. Writing from the margins tends to be more experimental and oppositional than writing from the center. Experimental writing, of course, is not owned by Jews. But when one is pushed into a corner and feels like a total outsider, experimental writing is the alternative because Jews have to live in alternative worlds. Gary Saul Morson (1996) explains,

> I recall that in the secular Jewish culture in which I was raised, it was thought to be characteristically Jewish to be both intellectual and, still more, heterodox. Of course, as the example of Chekhov demonstrates, the ideas of an intellectual who diverges even from the intellectual herd is not only Jewish, but, unless my upbringing was unique, it at least represents one important Jewish value. (p. 95)

Thanks to William F. Pinar, the father of what is called the Reconceptualization of Curriculum Movement that began in the 1970s, the field of curriculum studies is one in which scholars can do heterodox, experimental work. Pinar opened the field to many scholars who do not quite fit in to other disciplines. Pinar (1998) suggests that curriculum scholars remain wedded to "open intellectual exploration" (p. 3). The reconceptualized field has been an incredible experimental journey for many educationalists. Curriculum studies creates avenues especially for minority cultures to do the work of social justice, studying the complex lives of teachers and the subjectivities of students as related to larger sociocultural and historical movements. *Understanding curriculum*, as Pinar (2004) suggests, is a form of social psychoanalytic, racial, gendered, and historical text. These various theoretical constructs allow scholars to better understand lived experience, the *lebenswelt*, while delving deeper into terrain that turns on different social and psychic registers (Pinar *et al.*, 1995). Curriculum studies is the discipline that allows the Other to be Other, that allows one to write curriculum in cutting-edge ways. Curriculum theory allows scholars to find out who they are in the larger sociopolitical context. Curriculum theory opens up scholarly sites for the Othered to find a place to speak, a place to find a voice.

Following the work of Pinar (1994), I argue that identity-formation is historical, social-psychological, and cultural. As Daniel and Jonathan Boyarin (1997) point out,

> One of our most important ambitious goals is thus to shape a space of common discourse between Jews and others who share a critical approach to the politics of culture. By a "critical approach," we mean an understanding of history and identity that fully respects the powerful ways that they inform each other, yet also understands that in exploring and articulating our various identities we are simultaneously remaking history. (p. 26)

The only way to understand identity-formation is by studying historically. Thus, it becomes important to look to our ancestors to figure out who we are. Mortimer Ostow (1997c) contends that children who identify with parents become "interested in the concept of ancestors" (p. 164). Studying ancestors, for most Jews, is difficult, because our histories have been lost, family trees have been decimated. We are a people of the Diaspora and the exile. The scattering of Jews has created a vacuum of memory. Yet Jews continue to write about the experience of exile to try to capture what it feels like to be lost, to not be at home. Dorota Glowacka (1999) suggests that Jacques Derrida makes interesting connections between exile and writing.

> In his essay "Edmund Jabes and the Question of the Book," Jacques Derrida points to the link between a writer's fascination with the textuality of the world and the Judaic tradition. Derrida speaks of Judaism as "the birth and passion of writing," originating itself in the narrative of its exile. Derrida emphasizes the founding exchange between Jews and

> writing, the common root enhanced by the narrative of the Jews as wanderers in foreign lands, always displaced in relation to the place of their birth and never able to return home. (p. 100)

Because Jews live in cultures that ostracize and shun difference, there is never a home to which we can return. Writing about feeling exiled is a state of mind, rather than a literal place of exile.

American Jews might take comfort in studying the lives of European Jewish intellectuals because we are kin, albeit at a distance. As Sander Gilman (1999) contends, "Identity is what you imagine yourself and the Other to be: history/historiography is the writing of the narratives of that difference" (p. 2). This book project is an exploration of identity-formation, Otherness, and difference.

American scholars who are Jewish might look to their European Jewish intellectual ancestors and pay homage to them because they have made it possible for us to write as Jews and to write in such a way that we can better understand how we are situated in the world. Of course, situatedness is complex. David Simpson (2002) explains, "For now I want to register situatedness as a slippery term, whose slipperiness I hope to expound. I take it to designate an instability or obscurity describing our way of being in the world" (p. 20). Understanding the ways in which Jewish intellectuals situate themselves is not only important psychologically but is also important politically and socially. Situating oneself as a scholar allows one to be more in touch with one's psychic experience. Scholars do not write out of a vacuum, but write out of a specific historicity. This historicity needs to be addressed if one's scholarship is to be responsible.

Jewish Strangers

Georg Groddeck (1917/1977) remarks that "the connection between choice of profession and the unconscious have been pointed out" (p. 39). One might become a teacher and scholar because of the love for books. But there is more. If it is the case that one is driven to a certain profession by the "It," as Groddeck teaches strangeness must be involved in one's choice of profession because it is the unconscious that picks what it is that we do. Of course, I am not the first person to address these issues. Maxine Greene (1973) did this many years ago in her book *Teacher as stranger*. But my point is not existential, as Greene's was. My point is not abstract, my point is more psychological, more phenomenological. I want to find out about this strange feeling by examining what it is that Jewish intellectuals tell us about how they have felt. And although our professions are chosen perhaps unconsciously, at one level we know what we are doing because we work to understand what we are saying. Though much of our lives are driven by an "It," or an unconscious. However, one can still think through issues on emotional and intellectual registers. Human beings are not utterly without rational conscious perspective, no matter how clouded over by unconscious forces. One remains thoughtful and vigilant, even in the realm of the vague, the unconscious, the "It." One must never become less vigilant or less thoughtful because of the difficulties of understanding one's *lebenswelt.* Hannah Arendt (1958/1998) contends,

> This, obviously, is a matter of thought, and thoughtlessness—the heedless recklessness or hopeless confusion or complacent repetition of "truths" which have become trivial and empty—seems to me among the outstanding characteristics of our time. What I propose, therefore, is very simple: it is nothing more than to think what we are doing. (p. 5)

To think what we are doing against a backdrop of thoughtlessness—against a backdrop of unconscious fog—is the ongoing struggle of intellectual life. Many of the Jewish intellectuals I examine became creative thinkers partly because they felt alienated, they felt like strangers. They did not have to become intellectuals, but they did. I suggest that they wrote what they did partly because they felt estranged from the world. Their work would have been different had they been squarely in the mainstream. There is something about living on the margins that creates a certain kind of thought. Thought that springs from a marginal place differs from mainstream thinking. I am not suggesting that intellectuals who are in the mainstream cannot do good work or cannot be creative. Rather, my contention is that those of us who are on the margins produce a certain kind of work because we are on the margins.

Jewish Intellectual Companions

An interesting psychic interaction happens when one examines the lives of others while studying the self. As William F. Pinar and Anne E. Pautz (1998) point out,

> And so biography intertwines with the life history of the writer to reveal aspects of both the writer and the subject—different people whose

> merged and separated voices collaborate to form a complex text. (pp. 67–68)

Examining the life histories of Jewish intellectuals does indeed intermingle with that of the person who is doing the intellectual work. And this fact is strange. This study is a montage of strangeness. There is something deeply psychological in this montage of strangeness, something deeply resonating. These complicated lives are meant to form a complex text. What is astonishing is that many Jewish intellectuals have expressed a sense of alienation and strangeness. I am sure that this sense of strangeness is due to their Jewishness, whether they call themselves ethnic or religious Jews. Louise DeSalvo (1998) speaks to "Aspiring educational biographers" (p. 269) as they engage in life history work and she says that,

> if it is your predilection, you unashamedly bring whatever you are concerned with in your life into the arena of your works to enrich and deepen your understanding of someone else's life, to create an ongoing dialectic between your life and your work in which your work enables you to understand your subject's life better. (p. 269)

It is this movement of psychic dialectic that is of interest. Studying the words of others does not mean getting a transparent meaning. The psychological feeling of resonance is an interesting phenomenon when reading Jewish intellectual ancestors, of feeling that someone else feels similar feelings to one's own. To psychologically identify with Jewish intellectual ancestors might allow one to better identify with one's own pain of alienation.

Alan Block (1999) speaks to the issue of strangeness. He contends that "the Jewish voice positions the Jew as outsider, and even by the Jew him/herself, positions the Jew as stranger" (p. 164). Like Block, Paul Lauter (1996) says that, "In the 1960s, I learned to embrace my strangeness, Jewish and otherwise, because it allowed me to see what the conventions of America had been hard at work at hiding" (pp. 45–46). Though the situation in America has always been better for Jews than Europe, still many Americans are anti-Semitic. It only takes one downturn in the economy or a war to fuel anti-Semitism. Living in a country that is primarily Christian is not easy for Jews. Children especially feel ostracized at school if they are not Christian. Christmas time can be an alienating experience for Jewish children. Children begin to sense their Otherness at a very early age. Bad experiences in school reinforce Otherness.

Jacques Derrida (2003) in the film titled *Derrida* talks about his experiences as a Jewish child in school in Algeria when anti-Semitism was a state policy. He remarks that the trouble for the children had little to do with adults, but rather with other children. Jewish children were called "dirty Jews" by Gentile children. It is rather shocking that children can be so cruel. Derrida was kicked out of school—as a young child—because he was Jewish, during the Vichy years. This fact is also something hard to fathom. Why would *children* be expelled from school? Because of these experiences, Derrida says that he is extremely sensitive to any forms of racism. Perhaps this is why he is obsessed with the notion of the Other and notions of responsibility to the Other.

European Jews have always had it worse than Americans because anti-Semitism has always been more extreme in Europe. As I mentioned earlier, Austria and Germany have been particular hotbeds of anti-Semitism. Austria was perhaps the worst place for Jews before the Holocaust. But nearly every European country with the exception of Italy had a longstanding history of anti-Semitism. Thus, European Jews have always had to figure out complex psychological maneuvering in order to manage ongoing hostility. Perhaps Derrida's way of dealing with his early experiences of anti-Semitism shaped in some way his thought and work.

Psychologically, hatred forces people to hide, to split off a part of the self to protect the ego. As Sander Gilman (1995b) suggests, "It is this tension between the perception of belonging and yet not belonging to my given culture that mirrors the position of the acculturated Jews in late 19th century Europe" (p. xxvi). To belong and yet not belong is a complex psychological place to be. To be in and out of culture, to be in and out with one's Jewishness creates psychological tensions. To play *fort/da* psychologically with the given culture is tricky and tiresome. To play *fort/da* forces one to be better able to psychologically maneuver around potential disaster. Even still, when one is successful at maneuvering this difficult path, a sense of strangeness and anger persists. Unlike life at the margins, torments like these do not haunt people who live in the mainstream.

The following section pays homage to Jewish intellectual forefathers and fore-mothers in the form of a montage of words and worlds. This montage is not biographical in the same way that traditional biographies are. For one thing, I am not a biographer and I do not follow the strict methodology that biographers follow. The perspective here is from the field of curriculum studies. The focus here rests upon the complex relations between text, reader, culture, history, and education. I explore snapshot situations to show the ways in which Jews expressed their sense of alienation. I argue that intellectual work cannot be fully understood without taking into account more personal, psychological backdrops in which writers do their work. I think readers become less naive once they are clued into writers' situatedness. The contextualization of writers' lives is one key to why they wrote as they did. But still, the life history of people may tell us little about their writing in and of itself. Yet at the same time, paradoxically, life histories help to fill in certain gaps and give readers little clues here and there. Formalist readings of texts become, at any rate, more interesting against the backdrop of a more contextualized analysis.

A Montage of Jewish Intellectual Ancestors: School, University, and Beyond

The intellect is formed in school settings. Thus, this discussion begins by looking at autobiographies of Jewish youth who tell us the ways in which they experienced anti-Semitism in European schools. Again, this is not a comprehensive portrait, but rather a sort of montage. Let us start our examination of this problem with Jewish school children who did not grow up to be famous writers, like one Ludwik Stockel, a Jewish Polish boy born in 1914. Stockel reports that his school years were "tense." Not only that, his life at university was just as bad, if not worse because of the growing anti-Semitism in Eastern Europe before the Holocaust. Stockel (2002) reports that during his youth,

> [W]e avoided Catholics. Though the two groups were on speaking terms with each other and kept up appearances, the atmosphere [in school] was tense. Our mutual dislike was obvious; on the Catholics' side, it was further reinforced by the anti-Semitism of the priests who taught classes in religion. It was clear that any improvement in our relations with the rest of society would require a rational program of education, starting in the earliest years. (p. 172)

The disturbing passage here concerns the "anti-Semitism of the priests" who taught school to these young children. Of course it is a well known fact that the Catholic Church turned its back on Jews during the Holocaust, as historian Susan Zuccotti (2002) has taught us. Zuccotti (2002) tells us that

> Adolf Hitler's rise to power in January 1933 evoked no papal admonition of the Nazis' vicious anti-Semitic program. On the contrary, as the

> new Fuhrer imposed his initial anti-Jewish measures during the first spring and summer Vatican representatives led by Eugene Pacelli, the future Pope Pius XII, successfully negotiated a concordat between the Holy See and Germany. When the Nazis announced comprehensive anti-Jewish measures at Nuremberg in September 1935, Pius XI expressed no disapproval. (p. 8)

Zuccotti (2002) also stresses,

> [I]n the autumn of 1943, Pius XII . . . said nothing when hundreds of Jews were massacred on the eastern front and in Croatia. He had remained silent when hundreds of thousands more were seized in Catholic Slovakia, Poland, France and Belgium, as well as in Protestant countries, in 1942 and 1943. He was supposed to be a moral spokesman for all people. (p. 168)

I have dealt in depth with religious anti-Semitism elsewhere (Morris, 2001), but for now it is enough to state the fact that most European Catholics and Protestants were anti-Semitic. Priests and teachers did not hide this fact, as is seen in the example above. The sad irony here—and maybe this is an obvious point—is that one would not think that priests and teachers would be the ones to espouse anti-Semitism. One might expect anti-Semitic sentiments from thugs or uneducated, rural backward types. But this just isn't the case. In fact, education matters little when it comes to prejudice and hatred. Professors can be just as anti-Semitic as thugs and rural backward types.

Another youngster called "The Stormer" talks about his school experience in Poland before the Holocaust. This child talks about his intellectual formation in school, but he also talks about how anti-Semitic teachers put him down and belittled him. "The Stormer" (2002), states,

> School had given me a great deal, and it served as the basis of my subsequent self-education. In school I had also made many friends, who left indelible impressions on my psyche. From my school years I also recall a whole gallery of teachers of various types. One teacher, Mr. Fachalczyk, sticks in my memory. He caused me a great deal of trouble and used to make fun of my *peyes* and my long coat. He taught history and Polish. Today he is the leader of the Endek Party in our area and is known to be very anti-Semitic. (p. 236)

What kind of "history" would this anti-Semite teach? One hates to speculate. More to the point this "teacher" did not even try to mask his anti-Semitism. People like this do not even deserve the title "teacher." Teaching should be a sacred task. Teaching means to lead one out, to give children the opportunities to find out what their talents are and to inspire them. Hatred of whatever sort is NOT inspiring of course. Teaching is an ethical task, a profession that is tied closely to the notion of responsibility and care. Teaching hatred is not "teaching" at all. How does one educate young people while putting them down and "making fun" of them? On one level this "making fun" might seem harmless, but years later this is what "sticks" in this child's memory. Psychoanalyst, W. R. D. Fairbairn (1954) argues that the psyche only introjects the negative. Here is a good case in point. The negative memory—of being made fun of—that "sticks" with this child is the negative. Once the negative is introjected into the psyche it becomes permanently installed there, according to Fairbairn. If this child does not develop a good sense of self-esteem or is not in some way ego-syntonic, self-doubt can slip into self-hatred, which is a common phenomenon with oppressed groups. Negative introjects become permanently installed.

Another youngster, one E. M. Tepa—who also lived during the early part of 1900s and attended school in Poland—reports on his university experience. Tepa (2002) states,

> Moving from the care of my "ordinary" teachers in public school to the tutelage of the "extraordinary" professors here magnified my sense of self-worth. The anti-Semitic remarks that some of the professors made right in the middle of the examination, as well as other insults from my potential classmates, produced a rather negative picture of my future life in school. (p. 286)

Again, the disturbing thing in this passage is the audaciousness of the professor who publicly abuses his Jewish students. The professors' anti-Semitism is right there out in the open, out in the classroom for all the students to witness. No wonder children thought it was okay to name-call and engage in physical violence.

The above examples of the priest, the school teacher, and the professor show that many Poles did not try to hide their hatred, it was all out there for everyone to see. Americans probably find this rather shocking. American Jews—for the most part—have not experienced such bold and brazen anti-Semitism. This is not to say that America is not anti-Semitic; many sectors are indeed. Perhaps American Gentiles are less apt to publicly espouse this kind of hatred. American Gentiles hide their hatreds better than Europeans.

The Polish poet, Aleksander Wat (1988) reports that anti-Semitic comments were not reserved for high schoolers or university students. Here, in a disturbing statement, Wat (1988) says,

> But, for example, Andrzej, who was *six years old* [emphasis mine] at the time experienced anti-Semitism. I was bent on his going to an ordinary public school, not some exclusive school. I knew the principal of the school on Nowy Swiat, and so I sent Andrzej there. Ola went with him the first day. Andrzej didn't look Jewish except for his frizzy hair. The teacher wasn't in the classroom yet when the door closed behind Andrzej. But Ola had an inkling of trouble. She opened the door halfway and saw Andrzej up against the wall. He was surrounded by a bunch of rough kids who were shouting "You Kike!" at him and about to beat him up! There was plenty of anti-Semitism like that. (pp. 90–91)

What is absolutely unbelievable here is that a six-year-old child would be brutalized by other children! Although the ages of the other children who bullied this young six-year-old are not known, it is still highly disturbing that children would bully and beat up six-year-olds!! The indoctrination of anti-Semitism in children begins very early. This indoctrination happens not only at home, but at school and at church. And as we know from much historiography on the Holocaust, *children* were the first to be annihilated in the death chambers. The Poles and the Germans had no great love for Jewish children. The French under the Vichy regime sent some 6,000 Jewish children to Auschwitz. This even surprised the Germans (Morris, 2001).

Anti-Semitic incidents at school and university were also reported by well-known intellectuals like Martin Buber and Sigmund Freud, for example. What is striking about Martin Buber's (1948/1976) experience in school is that the encounter with Christian imperialism (read anti-Judaism) shaped his ideas on what he calls the "ghosts" (1948/1976, p. 168) of anti-Semitism. These so-called ghosts shaped his thinking about being Jewish in a Christian society. Buber (1948/1976) tells us about his school experience.

> The teacher and the Polish students crossed themselves; he spoke the Trinity formula, and they prayed aloud together. Until one might sit down again and we Jews stood silent and unmoving, our eyes glued to the floor. . . . [T]he obligatory daily standing in the room resounding with the strange service affected me worse than an act of intolerance could have affected me. (p. 29)

Buber writes about this particular experience many years later. It must have had a longstanding detrimental effect on him emotionally to remember it all those years later. Against the backdrop of anti-Semitism, Buber must have felt alienated and isolated as a child. For Jewish children, the issue of school prayer is not to be taken lightly. What Buber experienced as a school child is not that much different from what many Jewish children in public schools in America confront today, especially in the South where school prayer is prevalent. How is a Jewish child supposed to feel when Gentile children break out in prayer? Why is it that the Christian right cannot see the problem here? But then, the Christian right are not that dissimilar to the right wingers in Eastern and Central Europe.

Like Buber, Sigmund Freud expresses his discontent with the anti-Semitism prevalent in learning institutions. Freud (1935/1989), in his autobiography, expresses how isolated and alienated he felt as a Jewish intellectual working in an Austrian university. It is ironic that it was at the University of Vienna—and not, say, in Austrian coffee houses or just out walking the streets—that he felt most alienated. It is clear that many European universities were hotbeds of anti-Semitism. Freud (1935/1989) tells us that,

> When, in 1873, I first joined the university, I experienced some appreciable disappointments. Above all, I found that I was expected to feel myself inferior and an alien because I was a Jew. I refused absolutely to do the first of these things. I have never been able to see why I should feel ashamed. (p. 7)

When one experiences prejudice and discrimination, one is forced psychologically to deal with one's sense of self, ethnicity, and so forth. Peter Gay (1998) points out that it was no accident that Freud felt more keenly aware of his Jewishness while at the university. Self-formation gets shaped against a hostile atmosphere. Gay (1998) contends that "It was not without reason that Freud should date his particular Jewish self-awareness to his years at the University, where he began his studies in the fall of 1873" (p. 15). Moshe Gresser (1994) reports that when Freud became a faculty member at the University of Vienna, his promotion to professor was blocked because of anti-Semitism. Imagine not promoting Freud to professor!! At any rate, ongoing struggle with the university bothered Freud so much that he even considered giving up teaching.

Anti-Semites were not just thugs as I mentioned earlier. They were also university professors and some were even very well-known intellectuals. I have written extensively on the Heidegger affair (2001;2002) so I will not repeat that here. The question is not who was an anti-Semite, but who was not. What European Gentile intellectual was NOT anti-Semitic before and during and even after the Holocaust? Upon reading a Jewish Romanian playwright named Mihail Sebastian (2000) we learn about his friend, the well-known religious scholar Mircea Eliade. Sebastian tells us that Eliade was an anti-Semite. Eliade is a darling of Jungians because of his notion of the numinous. It is no secret that Jung was an anti-Semite too. At any rate, Sebastian (2000) reports disturbing encounters with Eliade and how their friendship eventually crumbled. Eliade told Sebastian that,

> "The Poles' resistance in Warsaw," said Mircea, "is a Jewish resistance. Only yids are capable of blackmail of putting women and children in the front line, to take advantage of the Germans' sense of scruple. The Germans have no interest in the destruction of Romania. Only a pro-German government can save us. A George/Bratianu/Nae Ionescu government is the only solution. The Soviets are no longer a danger. . . . What is happening on the frontier with Bukovina is a scandal, because new waves of Jews are flooding into the country. Rather than a Romania again invaded by Kikes, it would be better to have a German protectorate." (cited in Sebastian, 2000, p. 238)

How can someone so smart say something so stupid? Sebastian (2000), after hearing Eliade blather this garbage, says to himself "Just look at what he thinks, your ex-friend Mircea Eliade" (p. 239). Eliade knew full well that he was talking directly to a Jew, his friend, Mihail Sebastian, when he said these horrible things. Did Eliade not care that Sebastian was a Jew? Obviously not! This kind of talk is reminiscent of the encounter of Heidegger with one of his Jewish students, Karl Löweth. Knowing that Löweth was a Jew, Heidegger set up a meeting with him and wore, of all things, a swastika! (Morris, 2001). Did Heidegger not care that he was talking to a Jew? Obviously not!

Jewish intellectuals experienced all kinds of anti-Semitism in European society. Not only did they experience it in schools and universities but in everyday life as well. Lotte Kohler and Hans Saner (1992) tell us that Hannah Arendt experienced "feelings of being alien, homeless, and alone" (p. viii). It is interesting to note that Arendt turned to "literary companions" who also felt as though they were outsiders. Elisabeth Young-Bruehl (1982) says of Arendt that she had "friends of every sort and also the historical figures with whom Arendt felt special affinities, like Rosa Luxembourg and Rahel Varnhagen, had one characteristic in common: each was, in his or her own way, an outsider" (p. xv). The sense of being an outsider influenced Arendt's work. Arendt took a political turn because of external pressures she felt as a Jew in an anti-Semitic culture. At the time Arendt was writing, many philosophers had little interest in doing political philosophy. Elisabeth Young-Bruehl (1982) tells us that Arendt "had come to her political awakening not as a leftist but as a Jew" (p. 105). Arendt talks of being shunned for writing politically and accuses philosophical and academic communities of shying away from world events and politics, especially as the Nazis became a real threat. Arendt was enraged by the lack of concern or indifference on the part of her colleagues to the rise of the Third Reich. She was disgusted by their naiveté and apolitical sensibilities. Elisabeth Young-Bruehl (1982) recounts a story told by Anne Mendelssohn Weil about Arendt's reaction to her friend's indifference to the political situation in Germany.

> Anne Mendelssohn Weil remembered meeting her on the street one day in 1932 and hearing her talk of emigration for the first time: the rising tide of anti-Semitism around her was making the prospect of staying in Germany less and less reasonable, she said. Anne was surprised, and answered that she had not experienced any drastic increase in hostility toward the Jews. Hannah Arendt looked at her friend in amazement, said sharply, "You're crazy!" and stomped off. (p. 98)

Arendt was not fond of the academy, to say the least. In a letter to Karl Jaspers, Arendt (1992) says,

> The attitude of this country toward death will never cease to shock us Europeans. The basic response when someone dies or when something goes irrevocably wrong is: Forget about it. That is, of course, only another expression of this country's fundamental anti-intellectualism, which, for certain special reasons is at its worst at Universities. (p. 30)

Not only was Arendt troubled by the anti-intellectual nature of the university, but she was also distraught by the way in which German universities aligned themselves with the Nazi Party. Arendt (1992) states to Jaspers that,

> that is the very reason why it is so bad that the universities "lost their dignity" in 1933. I don't know how one should go about rehabilitating their reputation, for they made themselves ridiculous. Denazification, important as it may be, is, after all, only a word, because the institution itself—worse yet the standing of the scholar—has become ridiculous. (p. 50)

"Ridiculous" is not a strong enough word to describe the universities' complicity with the Nazi Party. "Criminal" would be a better descriptor. The universities fired Jewish professors and replaced them with incompetents, kicked out Jewish students, banned Jewish books from the libraries. This was the meaning of "alignment." One wonders about the political ramifications of American universities aligning themselves with government policy like No Child Left Behind: criminal.

Arendt had little respect for armchair philosophy and for professors who were not actively engaged in the work of social justice. She wrote about action and thought, not just thought. Arendt believed that true philosophy was a philosophy of activity, of doing. She argued that political activity was a public duty and responsibility. Dana Villa (1996) tells us that Arendt,

> rejects as unpolitical any conception of deliberative politics that desires to replace the "bright light" of the public realm with the more controllable illumination of the seminar room. For Arendt, to appear in public—to engage in political action—is necessary to perform. (p. 73)

Arendt draws a wedge between the sheltered world of the academy and the larger sociopolitical world outside. This, however, is a mistaken assumption. The public school or public university is part and parcel of the outside world, not cut off from it. Teaching is a public act indeed. Teaching changes the public, or at least it should. Teaching is the work of social justice. Teaching teachers is especially political because school teachers will take back to their classrooms ideas learned at the university that could change the sociopolitical world. Transmitting knowledge from one generation to another is a profoundly political act. As Arendt (1958/1998) says, "with word and deed we insert ourselves into the human world" (pp. 176–177). Is this not what teachers do? We speak words and perform deeds in the public space of the academy. Students are our companions in the public space of the classroom. Students and teachers talk together about what it means to be an educated citizen in the world of crisis and chaos. Today, we engage in serious discussion about who we think we are to one another and how we wish our children to grow up against the backdrop of 9-11, terrorism, and war. Now more than ever, Americans must seriously engage children in political discussions. Arendt (1958/1998) talks about the nature of the polis.

> The polis, properly speaking, is not the city-state in its physical location; it is the organization of people as it arises out of acting and speaking together, and its true space lies between people living together. (p. 198)

The classroom, I argue, is the polis; it is the place where academics live together with students. The classroom is the space where we live together as companions grappling with serious political, cultural, socio-psychological issues that affect teachers and students alike. What lessons are learned in the classroom can have tremendous impact on the next generation. If we are doing our jobs, children of the next generation will fight for social justice and equity for all peoples and fight for the right to take back the educative arena from the grips of the government. Standardization is "alignment." It is a crime.

Like Hannah Arendt, novelist Stephan Zweig expresses a sense of alienation in his autobiography as he discusses what it was like to live in Austria at the *Fin-De-Siècle*. In *The World of Yesterday*, Zweig (1943/1964) talks about what it was like to grow up as a Jew in Austria, in a highly anti-Semitic culture.

> I was born in 1881 in a great and mighty empire, in the monarchy of the Habsburgs. But do not look for it on the map, it has been swept away without a trace. I grew up in Vienna, the two-thousand-year-old super-national metropolis, and was forced to leave it like a criminal before it was degraded to a German provincial city. My literary work, in the language in which I wrote it, was burned to ashes in the same land where my books made friends of millions of readers. And so I belong nowhere, and everywhere am a stranger. (p. xviii)

It is a well-documented fact that what Nazis did was burned books. Are books that powerful? Obviously they are. Who would think that books would be *feared* so much that they would have to be burned. What does the burning accomplish? What does it mean to burn? What does it mean to have someone's words turn to ashes? Killing words is a symbolic killing of the author of those words. Zweig says "I was in a torment of anxiety" (p. 400). Can a writer today imagine her books being burned? How could one *not* be tormented by the fact that one's writings are being destroyed. Ones words are an extension of the self. When you kill someone's words, you kill the self as well. People are fearful of the written word.

The power of words. Dwell on this for a moment. Yes, words are not only powerful but can serve as psychic weapons against oppressors. As Sander Gilman (1995a) points out the "world of words" served to protect Kafka from the cruelties of anti-Semites in Prague. Through words one fights back. One creates interior worlds, flights into fantasy. Sometimes fantasy worlds serve as escape mechanisms to thwart pain. Through words, writers combat the terror of a hostile world. Words are expressions that help one psychically get rid of terror—at least temporarily—keeping ugliness from seeping too deeply into the psyche. Hostility does, though, have a way of seeping in. But continual writing, studying, and thinking helps one think through things and find passages out. Flights into the intellect serve as defensive shields against hatred and feelings of alienation.

Like Stephan Zweig, novelist Joseph Roth (1932/1995; 2002a, 2002b, 2002c) writes about what it is like to feel alienated, what it feels like to be the Jewish stranger. Interestingly enough, Michael Hofmann (2002) comments that Roth writes as if always on the verge of catastrophe. Strangerhood is being always already in a state of disaster. Strangeness symbolizes a world that is always already out of order. "Contemplating Roth's speed, his fatalistic interest in the overthrow of a character, his huge appetite for catastrophe . . . [marked his writings]" (Hofmann, 2002, p.10). Being Jewish, especially during the years in which Roth lived, meant living in a continual catastrophe. Living in a country that is highly anti-Semitic is catastrophic both psychically and physically.

There are all sorts of ways of thinking about what catastrophe might mean. Scapegoating Jews is a catastrophe. And this never seems to stop. With the war in Iraq, many blame the Jews. This is a catastrophe. It is a catastrophe to belong and yet not belong to a culture. In Germany, for example, Jews have always had a strained relationship with this country. Jews were not considered citizens of Germany, they were not even considered human beings during the Third Reich. Today there are many debates about whether to call Jews in Germany, German Jews or Jewish Germans. These significations may seem meaningless to an outsider, but for German Jews—naming—becomes an issue (Morris, 2001). If called German Jews, national identity is forefronted; if called Jewish Germans, ethnic identity is forefronted. Jews still have an ambivalent relationship to Germany. This will not change until Germany fully deals with its past, which it cannot do unless it emotionally works through the Holocaust (Morris, 2001). It is a symbolic catastrophe to continually attempt to maneuver complex psychic defenses such as these.

In a piece titled, "This Morning, A Letter Arrived . . ." Austrian novelist Joesph Roth's (2002c) character says,

> Now I was born nowhere and belong nowhere. It's a strange and terrible thing, and I seem to myself like a dream, without roots and without purpose, with no beginning and no end, coming and going and not knowing whither or why. It's the same with my compatriots, too. They

> have been scattered all over the wide world, they grip foreign soil with their frail roots, lie buried in foreign soil. (p. 167)

Whether one is literally exiled, as many Jews were before and during the Holocaust, or whether one feels that one must become psychically exiled and split off from the self, the feeling in both cases is lostness. The feeling of being split between a host culture and the Jewish culture is schizophrenic. Never being wholly grounded is, to say the least, a postmodern condition. But for Jews this postmodern condition takes on a unique flavor. The scattering of the Jews, or the Diaspora, is a particular historical, phenomenological and psychological issue that cannot be compared to other exiled groups. The Jewish stranger is not welcomed; the Jewish stranger is not wanted. The Jewish stranger is not at home anywhere in the world. And this estrangement is particular to Jews because of its particular historicity. This estrangement intensely impacts the psyche. Alan Bance (1932/1995) comments that in Roth's novel titled *The Radetzky march*, characters tried to fool themselves into thinking that the world was not falling to pieces, that the Habsburg Empire was not crumbling. "While deluding themselves that they are still at home in an intact world, they are already living in a kind of exile, in a dreamlike state" (p. xviii). Roth's short stories talk about a "dreamlike state" of coming and going, of being neither here nor there, of being in a state of limbo or a state that is not quite grounded. That is what exile often feels like psychologically. A limbo-like state is conducive to much anxiety. Neither here nor there, neither up nor down, neither over or across. Perpetual motion to nowhere. Limbo is vertigo. Living in a perpetual state of exile sickens. Roth's work is sometimes criticized for being nostalgic. Jewish characters in Roth's fiction long for the days of the Habsburg Empire—especially after it crumbled. The Habsburg Empire troubled though it was, protected Jews and other minority groups from persecution. However, Roth's nostalgic longing for the golden age of the empire when Jews were protected by Franz Joseph might be read as reactionary. Bance (1932/1995) explains,

> The sense of not being "at home in the interpreted world" in Rilke's famous phrase, is very much an Austrian one, and in Roth's case, leads to a constant reassuring reference back to his roots, as well as to a cult of the simple old Emperor as transcendent symbol of the better Austria. (p. xvi)

Austrian Jews did have it better under the protection of Franz Joseph, but some, like Freud, were not unhappy when the empire crumbled.

The empire was highly rigid and politically conservative, to say the least. The empire was oppressive and stifling. Yet, during the reign of Franz Joseph minorities like the Jews, were protected. It is easy to understand why—though—Roth grew nostalgic. Jews had little protection under the law throughout European history. It is ironic that under the iron fist of the empire, Jews were protected while under the rule of democracy in Germany, Jews were annihilated. Roth had his own theories about why the Nazis came to be. He (1996) states thus in his autobiography:

> If you want to understand the burning of the books, you must understand that the current Third Reich is a logical extension of the Prussian empire of Bismark and the Hollenzollerns, and not any sort of reaction to the poor German republic with its feeble German Democrats and Social Democrats. Prussia, the ruler of Germany, was always an enemy of the intellect, of books, of the Book of Books. . . . Hitler's Third Reich is only so alarming to the rest of Europe because it sets itself to put into action what was always the Prussian project anyway: to burn the books, to murder the Jews, and to revise Christianity. (p. 210)

This disturbing passage shows that Roth was well versed in history, even though his claims are rather exaggerated. It probably was not "always the Prussian project . . . to *murde*r Jews." I would imagine that many historians would argue with Roth on this one. Maybe many Prussians would have rather that Jews emigrated. But murdered? This thesis is only upheld by one Daniel Goldhagen, a contemporary historian who has been roundly criticized for this view. Even if it were the case that Prussians wanted to murder all the Jews even before Hitler arrived on the scene, there is no way to prove this claim. Nonetheless, Roth did not hold his tongue—he was deeply disturbed by the events at hand—and he shows how deeply engrained anti-Semitism was in Prussian society. This is the point that has the most validity. The sweep and depth of anti-Semitism in most European countries over history is a well-documented fact. One could trace anti-Semitism back to the early Church Fathers, or what is called the Patristic literature (Morris, 2001).

Unlike Joseph Roth, Franz Kafka was not nostalgic. Kafka was a critic of the vast bureaucracy of the Habsburg Empire. Kafka's (1958) story *The castle* is based on a real Castle that is located right outside of the central district in Prague. The Castle is a monstrous structure that sits on top of a hill and can be seen from all over Prague. The Castle must be spoken of with a capital C because of its massiveness. The C signifies—for our purposes here—monstrosity. Lower case C does not capture the pervasive ugliness and horror that is the Castle. The Castle, for Kafka, is a symbol of the evils of bureaucracy. Even today the Castle is ominous and spooky. Americans might have little understanding of the massiveness of the building and its grounds, for there is nothing comparable on American soil. There is something at once hideous, yet beautiful about the Castle. The Prague Castle—Kafka's Castle—reminds one of something one might read about in one of Edgar Allen Poe's stories or something one might see in Vincent Price movies. The Castle is a huge labyrinth, a town in and of itself. Kafka actually worked there as a clerk in a small office inside of the Castle walls. It is hard to find the way out once inside the winding pathways of the Castle walls. It takes *several days* to cover all the ground inside the pathways. If one has not been there, it becomes hard to imagine such a monstrous, gray, ominous, inhuman place. A day of touring the castle is not enough to see the whole thing. Some paths inside the Castle walls lead to gardens; other paths lead to prison cells underground. Some paths lead to buildings within the Castle walls; other paths lead to the outer parameters of the Castle. There is something eternally gray and depressing about the Castle. It is as if Dracula is lurking somewhere or vampires are waiting to pounce on victims—especially at night—inside the Castle walls. One might imagine bats hovering the outer parameters.

Kafka's (1958) novel *The Castle* could be a metaphor for the ominousness of Franz Joseph's reign, or maybe the Castle could be understood—broadly—as a symbol of what was to come after Franz Joseph. More than likely, however, The Castle was symbolic of Franz Joseph's reign. There was something terrifying about the control and orderliness of the empire; it seemed that nothing ever changed. Everyone was under the eye of a huge, roving, intricately bureaucratic panoptican. Kafka's (1958) satirical gaze is evident early on in *The Castle* when the character "K" engages in a discussion with a teacher. What is significant for our purposes here is that K talks to a *teacher*. Why would K talk to a teacher, rather, than say an office worker?

> "You are looking at the Castle?" he asked more gently than K had expected, but with an inflection that denoted disproval of K's occupation. "Yes," said K. "I am a stranger here, I came to the village last night." "You don't like the Castle?" asked the teacher quickly. "What?" countered K., a little taken aback, and repeated the question in a modified form. "Do I like the Castle? Why do you assume that I don't like it?" "Strangers never do," said the teacher. (1958, p. 13)

Strangers never do like the Castle, says the teacher. The teacher is at once the symbol of anti-Semitic Prague. Strangers—Jews—do not like the provincial feeling of Prague, the anti-Semitism of Prague. The Castle is a symbol of hate that Kafka felt as Jew. The hatred of Jews in Prague was vast/heavy/pervasive, like the Castle. Kafka felt that teachers—petty bureaucrats and anti-Semites—were complicitis with everything that represented what was wrong with Prague. In other words, teacher types were enemy because in those days teachers were little more than autocrats, in fact many teachers acted in sadistic ways toward pupils, like dictators. This fact has been well documented in the historical record. Austrian school systems—which were not that different from the schools in Prague—were wretchedly autocratic. At any rate, Kafka's character—who is probably Kafka himself—does not feel part of the teacher's world. That is, Kafka—through—the mouthpiece of "K."—expresses his outsider status. Sander Gilman (1995a) explains:

> Kafka's membership in the Jewish minority in Prague, the third city of the Austro-Hungarian Empire; a father who had originally spoken Yiddish, then Czech, and only then German; and his employment in the new Czech Republic as the only ethnic "German" in a workmen's organization company—all these reinforced his sense of marginality. (p. 12)

Marginalization is always already problematic. More specifically, the marginalization of Yiddish disturbs. For taking away one's language erases one's identity. The shadow of Yiddish is what troubles. Yiddish is not considered a "real" language by anti-Semites. Because Germans, Austrians, and Czechs (who are not Jewish) do not understand Yiddish, it is considered a threatening secret code among Jews. Kafka found himself steeped in language and it was through language that he expressed anxiety. Kafka's anxious fantasy world gets its meaning from language, from words. But Kafka did not write in his home language, Yiddish. Kafka did not write in his *mameloshen* (mother tongue), rather he wrote in German. Now this is a significant fact. What does it mean to write in a language that is not one's own? What does it mean to write in the language of the oppressor? What does this do to the psyche? In some sense, it must split the psyche. Room must be made for the oppressive language while keeping the home language on the back burner. Yet when one makes room for the oppressive language what is one doing to one's soul? This fact becomes especially haunting when we know what happened to European Jews in the years following Kafka's death. Grace Feuerverger (2001) says,

> my relationship with Yiddish, my first language, my *mameloshen* . . . [was lost] . . . the sense of loss and mourning embedded within the symbolic

> meaning of Yiddish, the language spoken by millions of Eastern European Jews for nearly 1,000 years until the outbreak of WW II . . . its native speakers were systematically murdered by the Nazis. (p. 13)

The annihilation of the Jews was the annihilation of Yiddish as well. Killing language is killing people. Killing people is killing language. The annihilation of the name, the letter, the word, the expression, the phrase is the death of a people. The death of a people is the annihilation of the name, the word.

Kafka understood that his work had to be written in German, not Yiddish. Writing in a language that is not your own must cause psychic strain. Eastern European Jews like Kafka faced these kinds of problems unfamiliar to many American Jews—especially to those of us who belong to the third generation after Auschwitz. Most of us have little connection with Yiddish. It had already been long gone in the ongoing assimilation of American Jewry by the second generation of (American) Jews after Auschwitz. For reformed Jews this may be fact. Perhaps in orthodox families Yiddish remains. But the orthodox remain a minority in the American Jewish community. Yiddish fades with each passing generation, as most of us become more and more assimilated into American culture. This is troubling because with the passing of Yiddish, so too passes the language the Jews. With the passing of the language, part of our collective identity disappears.

Sander Gilman (1995a) points out that for Kafka, fantasy helped him endure constant pressures of an anti-Semitic culture. But that fantasy, because expressed—in and through words—became part of the public sphere. Fantasy, that is, becomes public once written down. What was once private, now is open to public scrutiny. One must be cautious about sharing private thoughts in public. Today, though, most private thoughts are made public because of our posthuman condition. Connected to high speed Internet—always already hooked into technology—consciousness is downloaded or uploaded, as it were, via computers. Email, the private letter writing to the Other is made public by the very fact of the Internet. What one writes to the Other is immediately open to inspection by the whole cyberspace world. This is a disturbing phenomenon. When writing an email, the observing Ego cautions and watches over a nervous writing body-ego-cyborg ex-pressing private thoughts that at once become wholly public. Thus, fantasy today no longer remains secret. There are no secrets in the posthuman culture.

Let us return to Kafka. Gilman (1995a) suggests that fantasy—like Kafka's—is embodied because it emanates from the "tongue" of the writer.

> The writer's tongue is but an extension of that phantasy body. As Deleuze and Guattari note: "Rich or poor, each language always implies a deterritorialization of the mouth, tongue, and the teeth" (p. 19). This is a tongue intimately connected, as Alexander Kluge and Oskar Negt have argued, with self-preservation in the public sphere. (p. 24)

Fantasy is more than embodied today. Fantasy is also cyberembodied. One cannot escape the fact that embodiment is forever altered by cybertechnologies. Writing is no longer done solely with the pen. One must wonder what cyberembodiment does to one's soul.

At any rate, back to Kafka. Some might read Kafka's *The Castle* as 481 pages of babble. The strange language of the Castle makes little sense as it winds around and around—like a vast labyrinth—the theme of alienation. This babble-like language symbolically describes a place of vast Otherness. And in this world, in this vast labyrinth, "K." feels "wretched" (p. 197). "Though for the moment K. was wretched and looked down on, yet in an almost unimaginable and distant future he would excel everybody. And it was just this absolutely distant future and the glorious developments that were to lead up to it that attracted Hans; that was why he was willing to accept K. even in his present state" (Kafka, 1958, p. 197). To live as a marginalized human being is to be looked down on by the majority culture. To be constantly humiliated and insulted is the way in which a minority culture must live. The question is how does one survive constant persecution? What part of one's psyche gives? Where does the madness begin to take hold? What part of the self becomes filled with self-hatred? What part is the fighting part? How much can one endure? How often does one think of suicide? Where does the self go to find refuge? For Kafka, his refuge was his writing.

To be a continual stranger in an ominous land is K.'s legacy. *The Castle* is really a story about the Jew as Stranger. This is not a new theme in Jewish literature, but one that repeats itself over and over again throughout Jewish history. Talia Pecker Berio (2002) comments that "There has always been a 'here' and an 'elsewhere' in Jewish history. The condition of otherness, the nonbelonging or half-belonging to a place, a land, a nation, was imposed from without and cherished from within" (p. 95). What remains a question here is whether Jews "cherish" alienation. Some might. Some might not. Some might sink in alienation and commit suicide. Others might make alienation useful and write out of a space of Otherness. Cherishing it, perhaps, is a bit of an exaggeration for many. Can alienation ever really be cherished? Alienation certainly produces a psychic tension for many and it was partly—one would venture to guess—this tension of belonging and not belonging that inspired Kafka to write what he did. Whether or not he "cherished" these feelings remains a question. Perhaps he would not have written what he did if he were not Jewish and marginalized.

A writer who is often compared to Kafka is Bruno Schulz. Schulz's writings are surreal, phantasmagoric. His writing is not so much satirical as much as it is just weird. He tries to capture—from the perspective of the Jewish outsider—small-town life in the Eastern European provincial town of Drohobycz. Jerzy Ficowski (2003) remarks that Schulz "felt prey to a frightening strangeness that overwhelmed his very existence" (p. 127). Of special note for our readers who are teachers—Schulz was an art teacher. Not that he liked being a teacher, clearly he did not. In fact he felt tormented by teaching because he thought it was a waste of time. Writing and producing works of art interested Schulz much more than teaching, which he saw as drudgery.

Schulz is best known for his surrealism. In *The age of genius*, Schulz (1979) writes,

> The large colored picture painted on the front of the stove grew blood red; it puffed itself up like a turkey, and in the convulsions of its veins, sinews, and all its swollen anatomy, it seemed to be bursting open, trying to liberate itself with a piercing crowing scream. (p. 15)

Schulz's language bursts at each sentence, the words bloat. The words seem too vast, too overgrown. Everything in Schulz's writing is phantasmic, over-exaggerated like a dream. Jerzy Ficowski (2003) remarks that the "age of genius" refers to a psychic state whereby inside and outside are not clearly delineated. This phenomena is not dissimilar to dreamlike states described by psychoanalysts Naomi Rucker and Karen Lombardi (1998). They argue, "Mental life is not a structure but, rather, a mode of lived experience and a process of linking between inside and outside, self and other" (p. 22). Linking inside and outside, self and other, suggests fluid-linking capacities blurring boundaries between consciousness, unconsciousness, and the world at large. Hence, it becomes difficult to say—psychically—where the inside ends and the outside begins. When dreaming, confusion arises. Sometimes one dreams that one is dreaming yet it seems impossible to awaken. What does it mean to dream that one is dreaming? Or what does it mean to be awake and feel like one is still dreaming? This phenomenon happens at twilight moments of fatigue. Sometimes while one is awake one is not so sure if one dreamed something or whether that event which one dreamed really happened. It is all confusing and confused. Schulz's writings convey experiences like these. Ficowski (2003) says of Schulz,

> In accordance with its mythic model, Schulz describes his childhood as an "age of genius," a time when no barriers existed between the inner psyche and the outer world, between dream and reality, between desire and fulfillment, between the intellectual and the sensual—the time of the origins of poetry. (p. 72)

Schulz's sensitivity to perpetual movement of the inner and outer, to the fluidity of psychic life is uncanny. Whether or not his experience as a Jew in a hostile world allowed him to develop a heightened sense of awareness puzzles. However, whether or not marginalized people(s) are able to develop highly tuned uncanny modes of awareness—because of necessary retreat into phantasy-life—remains an open question. Schulz's writings are a good example of the phantasmic, the fluid, the bizarre, the dreamlike. He writes as if he were dreaming, and dreaming is a fluid, bizarre register. Lombardi (1998) argues that conscious and unconscious processes are continually in movement between inner and outer realms and in which, "transformations are bidirectional and dialectic; it is not that the unconscious is made conscious, that there is a continual interplay between conscious and unconscious processes" (p. 4). This continual interplay of conscious and unconscious is what gets expressed in Schulz's (1979) writings. But whether or not readers can translate the writings is yet another question. Interpretation of fantasy life in its most dreamlike state becomes guess work at best. "From hour to hour the visions became more crowded, bottlenecks arose, until one day all roads and byways swarmed with processions and the whole land was divided by meandering or marching columns-endless pilgrimages of beasts and animals. . . . My room was the frontier and the tollgate. Here they stopped, tightly packed, bleating imploringly" (Schulz, 1979, p. 17). Translation? The translation must always already be Other. Altered states lead to Alterity-in-translation. Perhaps this passage is beyond translation. Hallucinatory states perhaps remain just that. What to make of the bizarre? Perhaps the question is, why hallucination? What purpose does hallucination serve? These questions will be addressed further on in the book, but for now let us simply raise the question.

Let us leave the question unanswered. Some questions must remain questions. That is what Otherness is. It is a question, a remainder, a difference that has no same.

Interestingly enough, Schulz read his dreamlike writings to his students to inspire. Jerzy Ficowski (2003) contends that Schulz's ability to get through to his students had a lot to do with the bizarre tales he would tell. Students' interest piqued in response to the bizarre. In fact, Ficowski thinks that the reason Schulz was beloved as a teacher is because his teaching seemed mythical. Ficowski (2003) comments,

> When Schulz wanted to win over the class and inject some variety into the lessons, he customarily used a foolproof method, especially in the lower grades: the telling of fairy tales. There is no one among the former students of the Drohobycz gymnasium who does not recall these extraordinary classes during which one forgot about school, about everything, and listened with bated breath to the strange tales told by Professor Schulz. (p. 49)

Bruno Bettelheim (1989a), in his well-known book titled *The uses of enchantment*, argues that fairy tales (like the mythical stories Schulz told to his students) connect with unconscious stirrings and allow children to work through repressed material. It is not surprising that Schulz was able to connect with his students. There is something about myth and fairy tale that allow children to think otherwise, that allow children to imagine difference, to imagine strangeness. "Strange monsters, question-mark apparitions, blueprint creatures appeared, and I had to scream and wave my hands to chase them away" (Schulz, 1979, p. 17). John Updike, commenting on Schulz's work says,

> The pages are crowded with verbal brilliance, like Schulz's brimming, menacing, amazing skies. But something cruel lurks behind this beauty, bound up with it—the cruelty of myth. Like dreams, myths are a shorthand whose compressions occur without the friction of resistance that reality always presents to pain. In his treasured, detested loneliness Schulz brooded upon his past with the weight of generations; how grandly he succeeded can be felt in the dread with which we read even his most lyrical and humorous passages. . . . Something alien may break through. (Updike, intro in Schultz, 1979, p. xiv)

The cruelty about which Updike speaks—is to come—after the fall of the Habsburg Empire—with the rise of the Third Reich. Schulz uncannily sensed something horrid was coming; his writings are a strange foreboding of things-to-come. As Naomi Rucker and Karen Lombardi (1998) point out, "prescient knowledge" (p. 23) is not something one should dismiss out of hand. Schulz intuited that something terrible was happening, the future remained a question. In fact, he seemed to sense, like Joseph Roth, that life (for Jews) was catastrophic. Jerzy Ficowski (2003) comments that "Bruno Schulz's entire life was overshadowed by his intuition of imminent danger. Ordinary daily discomforts assumed the dimensions of catastrophes" (p. 127). Perhaps ordinary daily discomforts were felt more acutely by Jews than non-Jews. Being-on-edge forces one to become more tuned in to the negative. Danger loomed for Jews in Eastern Europe. Pogroms and ritual murder trials were common occurrences. There was reason for Schulz to be paranoid and depressed, filled with a sense of dread and foreboding.

We hear the sense of dread and foreboding in the musical compositions of Gustav Mahler, another Jew who was tormented with feelings of alienation. Leon Botstein (2002) says "Here was the artist as vulnerable individual, struggling with conflict and alienation" (p. 5). Mahler's symphonies are vast epic-dramas that sweep the listener along in painful, unresolved minor keys. The weight and heaviness of Mahler's lyrical chromaticisms may move listeners to tears. We travel with him up and down the mountains of Austria and into the heart of darkness listening to his tortured, tormented symphonies. There is something inescapably painful about Mahler's music. Theodor Adorno (1992) eloquently states,

> Mahler's minor chords, disavowing the major triads, are masks of coming dissonances. But the impotent weeping that contracts in them, and is rebuked as sentimental because acknowledges impotence, dissolves the formula's rigidity, opens itself to the Other, whose unattainability induces weeping. (p. 26)

The Other in Mahler can never be reached. The Other, Adorno suggests, is Mahler himself. "Mahler is the Other" (1992, p. 14). Not only was Mahler the other, he was Othered by Austrian society. Not unlike Freud, Austrians took little interest in Mahler's work—while he was alive. Mahler's music was not appreciated during his lifetime, in his home country of Austria. In fact, Mahler was shunned. Adorno claims that he was shunned because he was Jewish. Constantin Floros (2002) says "Mahler became the paradigm of the composer who failed to win recognition during his lifetime" (p. 13). He failed to win recognition because of anti-Semitism. The anti-Semites interrupted

Mahler's career and made it impossible for him to stay at the Vienna Opera House. Mahler was seen as a "rebel" Jew. Botstein (2002) explains,

> [T]he realist narratives embedded in symphonies form in Mahler's first three symphonies, when they contained direct references to nature, were, in his Nazi-era critics recognized, unstable and complex; they suggested fragmentation and rebellion. (p. 26)

The "rebel" Jew was hated by anti-Semites, even when he converted to Catholicism. Charles Maier (2002) tells us that

> Mahler took up his position at the Court Opera in the same year that Emperor Franz Joseph allowed Lueger, who had mobilized an anti-Semitic Christian Social electoral coalition, to take office as Mayor. Despite the conductor's nominal conversion to Catholicism, the City Council declined to accept a concert by the Opera orchestra if Mahler insisted on conducting it. (p. 68)

Anti-Semites did not want Mahler conducting anything, nor did they want to hear his so-called degenerate music. It is disturbing that today, when one tours the grand Opera House in Vienna, a huge portrait of Mahler hangs on the wall in the front entrance hall. Today, Mahler is Vienna's heroic son. Yet, the Viennese treated Mahler atrociously during his lifetime. Now, it seems he is memorialized. Is he memorialized out of a sense of guilt? Or do the Viennese have short memories? Or do they think people have short memories? Do they think that nobody really cares anyway. The display repulses, especially against the backdrop of the Holocaust. After Mahler left the Vienna Opera House for New York, all of the singers he hired were summarily fired. This should not surprise if one knows one's European history. Yet still it surprises, in fact it horrifies. Such blatant hatred, such blatant discrimination.

Feelings of ostracism and strangeness permeate Mahler's symphonies. No wonder. These feelings might be directly related to Mahler's Jewishness—one would imagine. Talia Pecker Berio (2002) comments,

> The feeling is that while Mahler's music cannot be described as properly Jewish, only an assimilated Jew of his time could have written it. But we'll never be able to say whether Mahler himself sensed it or thought of it deliberately. In his case we can speak of a virtual Jewish aesthetics: Whether it is conscious (and in that case entirely secret) or not, our discussion of it must remain essentially metaphoric. (p. 94)

A "Jewish aesthetic" is an interesting idea, however, it essentializes. One might say that there are as many "Jewish aesthetic(s)" as there are Jews. Jews respond to the world in unique ways, certainly Mahler's music is uniquely Mahler's. There is no other nineteenth century composer who comes close to Mahler in scope or depth. Mahler's music is an "aesthetic(s)" of pain and melancholy, grief and agony, suffering and loneliness, torment and anger. Adorno's (1992) claim is that Mahler's music creates a cathartic space in which to express melancholy. Adorno (1992) states,

> If, at the risk of nearly misunderstanding, one is attempted to compare Mahler and Stravinsky with trends in psychology, Stravinsky would be on the side of the Jungian archetypes, while the Enlightenment consciousness of Mahler's music recalls the cathartic method of the Freud who, a German-Bohemian Jew like Mahler and countering the latter at a critical point in his life, declined to cure his person out of respect for his work. (p. 39)

It is not certain what Adorno means here by the term "cure." Perhaps he is referring to the problem of "curing" Mahler's neurotic personality. But who is not neurotic? Adorno (1992) suggests, then, that Mahler needed his neurosis in order to create, perhaps his musical creations sprang from neurosis. Caution becomes necessary, though, when talking about Jews' creativity and neurosis as a form of madness. It has been long argued that Jews are *innately* mentally ill. But as Sander Gilman (1998) points out, this mythology serves to Other the other. Gilman (1998) explains,

> In the course of the 19th and early 20th century, a number of approaches were taken to the myth of the mental illness of the Jews. European biology, especially in Germany and France, served to reify accepted attitudes toward all marginalized groups, especially the Jews. The scientific "fact" that the Jew was predisposed to madness would have enabled society, acting as the legal arm of science, to deal with Jews as it dealt with the insane. However, the reality was quite different while the fantasy of the privileged group would have been to banish the Jews out of sight and into the asylum, the best it could do was to institutionalize the idea of the madness of the Jew. (p. 109)

Gilman's point demands careful attention. When talking about madness, it is important to understand the context in which this concept is generated. In the context of my study, I am not suggesting that Jews who work in universities go mad or are mad. That is not my argument. Although I am a Jew who works in a university, clearly I am not mad. What I am suggesting in my study is that Jewish intellectuals who work in universities feel ungrounded because the university symbolically feels like a madhouse since it has no aim. I argue that the university is no longer grounded in anything. It is hard to get a handle on what the university stands for anymore. Working in places that have little understanding of purpose feels chaotic and intensely Other. The university, I argue, is a site of dystopia where scholars no longer fully understand what they are doing inside of these institutions. A dystopic university becomes that site where people feel alienated and strange. A dystopic university becomes that postmodern space of chaos and uncertainty. A dystopic university creates a constant state of anxiety, especially for Jews who are always already marginalized. For Jews, the problem is twofold. For one thing, universities are *still* anti-Semitic places. Jews have to deal with this *and* deal with the fact that the university is like a madhouse.

A Sidebar: History, Psychoanalysis, and Jewish Intellectual Ancestors

To digress, let us think on a meta-level here—for a moment—about what it means to write about Jewish intellectuals historically as well as psychoanalytically. Writing about Jewish intellectuals who lived in the past, is ultimately writing historically. Re-creating lost realities is indeed healing; re-creating life histories of Jewish intellectuals teaches that one is not alone in suffering, that Jews have had a history of suffering. To better understand one's life against the backdrop of the historical archive, it becomes important to turn to these ancestors. It is through their sufferings that we (Jewish intellectuals) learn about our own. The learning process is painful. What we might gain is companionship—as distant and metaphorical as it may be. If anything, we learn that the life of many Jewish intellectuals is one of trauma. It is indeed traumatizing to be hated. This hatred comes in all forms. One does not need to dig very deep to understand anti-Semitic projections. Jean Amery (1999), in a piece titled "Anti-Semitism on the left—The respectable Anti-Semitism," comments,

> If I were cynical enough, I would quote the American mathematician and chansonnier Tom Lehrer, who already years ago on the occasion of the American "Brotherhood Week" sang: "And the Catholics hate the Protestants and the Protestants hate the Catholics and the Moslems hate the Hindus—and everybody hates the Jews." (p. 115)

Although this song is funny, it is—at the same time—not funny at all. It is not funny at all because it is true. Growing up in a Christian culture as a minority troubles. Many Christians tend to be insensitive and downright anti-Semitic. Moreover, the university is no haven for Jews—nor has it ever been. When a Jew is hired as a faculty member, some anti-Semitic faculty hold that Jews are already "over-represented" in the university. When "too many" Jewish students enroll, anti-Semites feel that Jews are "over-represented." When a Jew writes columns for a newspaper, anti- Semites argue the press is "controlled by the Jews." When a Jew directs a Hollywood film, anti-Semites say Jews "control the movie industry." Ah—A vast political conspiracy. Talking with one's hands is considered being "too Jewish." Being intellectual is considered to be "too Jewish." Elisabeth Freundlich (1999), in an autobiographical novel titled *The soul bird* says "I had often heard that one should not talk with one's hands because it is Jewish" (p. 143). Absurd. Having money is a "Jewish trait." Cheap. Now that's really Jewish. And so forth. The litany goes on and on. It is just disgusting. And what makes it really revolting is that these comments are contemporary. It is as if the Holocaust never happened. Everybody still hates the Jews. We just can't be hated enough. No amount of teaching multiculturalism and including Jewish literature will change hardened hearts, in fact, it might make them harder. One must wonder what it is about hatred that is so alluring. Hatred takes energy. Hatred takes time. Yes, it takes time to hate. It takes calculation and thought—or perhaps we should say thoughtlessness. Rural communities may be the worst places for Jews. Perhaps that is why many Jews have historically congregated in large cities. One must wonder what goes on in small backwater towns that actually promote hatred via billboards. Take for example highway I-80 in Georgia. Huge signs on the road read "Join the Son's of the Confederacy Now!!" The rural South has never been known to be welcoming to blacks or Jews for that matter. But hatred is not just a rural southern problem in the United States, it is everywhere.

Let us for a moment focus on the anti-Semitic slogan that Jews are "too intellectual." The irony is that in universities—especially ones where intellectual labor is not valued—to one's great surprise—Jewish faculty are often chastised for being "overly intellectual." Sander Gilman (1997) explains that "The Jew is certainly seen as 'over-intellectualized,' and this over-intellectualization is one of the sources of his pathology" (p. 52). Jewish intellectuals are suspect because they are seen as pathological. A pathological intellectual? Being intellectual means being diseased? Disturbingly, Gilman points out that Joseph

Goebbels contended that Jews were "too intellectual." Gilman (1997) says " 'intellectuals' quickly became the Nazi code word for the Jews, as it had been in France during the Dreyfus Affair. Thus at the book burning in Berlin on May 10 1933 Joseph Goebbels announced the end of 'an age of exaggerated Jewish intellectualism' " (p. 84). The historicity of talk like this disturbs, troubles. All of the slogans mentioned above are—in fact—deeply entrenched in Christian anti-Semitism. What does it mean when Christians say that they must "pray" for the Jews? Let us leave that an open question.

Contrarily Sander Gilman (1997) points out that Jews are simultaneously perceived by anti-Semites as not being "true intellectuals" but frauds, bloodsuckers, parasites, imitators. Jews are vampires, child killers, money grubbers—not "true" intellectuals. Jews are not "real intellectuals," they are second rate at best. Gilman (1997) says of Theodor Gomperz—who spouts such garbage—that

> "Jews are of the first rank only in the realms of the reproductive arts—acting and musical performance." . . . Jews may be visible as performers, but they are not creative in Wagner's sense and do not contribute much to the production of literature or music. (p. 107)

If Jews are *both* overly intellectual, but not really intellectual what are they? What is one to make of this contradiction? Psychoanalytically, hostility gets projected in contradictory ways. These contradictions are so extreme as to be curious; we simply cannot be hated enough. If Jews are first-rate musicians, why the hatred toward Mahler, why were all those Jewish musicians in his opera company summarily fired?

Because of anti-Semitic hatred, being Jewish means being traumatized. Living in a state of psychic exile one must psychologically negotiate between one traumatism(s) and another. Traumatism(s) eventually may empty the self of energy and vitality. Or, conversely, traumatism(s) enrage and energize. No matter, traumatism(s) alter the ego and force one's alter ego to respond. Responding responsibly means facing hatred head on, confronting it with full integrity. Yet every insult results in psychic wounds, every psychic wound travels in a sort of "migration," (Vansant, 2001, p. 12) to a split-off part of the self. Pain is continually housed in split-off parts of the psyche. Jacqueline Vansant (2001) remarks that

> Argentinean psychoanalysts Leon and Rebecca Grinsberg . . . say that "Migration as a traumatic experience comes under the heading of what have been called cumulative traumas and tension traumas, in which the

> subjects' reactions are not always expressed or visible, but the effects of such trauma run deep and last long." (p. 12)

Traumatic(s) traces leaves its mark on the psyche. But what exactly that mark is one might never know. The mark of trauma remains hidden, secret, repressed, pushed down, invisible. But what is invisible may—in time—become visible through melancholia, say, or shame.

Generally speaking, studying anti-Semitism can be traumatizing in itself on a different level. When confronted with hatred, the scholar cannot but stop short, the scholar must take pause. Taking pause means taking time to understand the ways in which one is being affected by one's studies. But one never really understands the full extent of what happens to the psyche when faced with the negative. The archive of Jewish intellectuals is not complete without confronting anti-Semitism head on. Here, the archival work symbolically becomes a site for trauma. There is little peace in working on the Jewish past. Making oneself ready for traumatism(s) is not possible. One cannot prepare—in advance—one's trauma. One can only be a witness to the psyche's upheavals on a very superficial level, for most of what gets wounded sinks into the invisible unconscious. Dreams might push up traumatism(s) in disguised forms. But still the translation of dreams only tells so much. The remainder and the trace of the wounds remain. No amount of psychoanalysis can heal the wounds of deep trauma. One always already lives in a state of (post)-traumatic stress. (Post) is not post. (Post) signifies the never ending, ongoing stream of trauma as it travels the length and height of the embodied soul for the entirety of one's life. The best one can do is witness and testify to the trauma through writing and other forms of expression. To make the invisible (trauma) visible to the Other. To make the Otherness visible. Otherness must be ex-pressed, pushed outward in some form of communication. Stories must be told. Healing, however, will never be fully possible. Or maybe at some level healing might happen. One never knows. Who knows. Whatever we do is not certain, but we must carry on the tradition of telling our stories, of teaching the truth about hatred and its impact. The Other must be heard and taken seriously. Otherness must be studied seriously and taken seriously. We are always already Other to each other. Alterity is a gift; alterity demands attention, tending toward, tending.

Working on Jewish intellectual otherness is always already fragmentary.

Jewish experiences of estrangement are vast. What is documented is vast, but one must wonder what is not documented? What sites of hurt, pain, humiliation go undocumented? What does it mean to suffer silently without ever becoming part of the historical record at hand? Another generation of Jewish scholars is yet to be born and let them too write their traumatic(s). Let the next generation document what needs to be documented and not forgotten. For what is forgotten comes back tenfold. To forget is to be buried in the ruins of psychic hell.

The purpose of this particular archive is memory of trauma, witness to injustice, cruelty and wrongs done to an entire peoples. This testament creates fragmentary openings. Spaces where one might learn a little about the ways in which entire peoples are humiliated still. What gets experienced confuses. Phenomenologically, felt experience is nearly incomprehensible and not-transparent. But scholarship on traumatism, on this particular traumatism, must work to bring forth some sense of understanding through the historical and psychological record; the archive of hatred must be known. Discontinuity, disjointedness, confusion, and chaos abound. But words give some kind of order to the chaos, words try to capture the uncapturable.

Intellectual Ancestors: Freud and Repetition

Sigmund Freud is perhaps the most written on, written about Jewish intellectual of our times. Freud's work and life are well known and well documented. But that does not mean that the writing on Freud can end; it can never end. In fact, it is never ending because Freud's work and life fascinate and continue to teach us about ourselves. What is of importance here is to look at his biographics to find out what conditions he worked under as Jew while living in a hostile place. Of course, it has been done and done by others. But the question raised here remains. How does a Jewish intellectual survive a hostile culture to produce world-class scholarship? This question has not been addressed enough. How much more can one read of Freud, or hear Freud's name? Can we bear any more work on Freud, or is this repetition more of the same. Is repetition always already uneducative, or is it in new ways educative? Freud and the repetition of his name signifies that we are still learning from him and from his example. This study is a testament to just that. It is indeed educative to study, study and study again our intellectual ancestors so that we may learn more about our own intellectual contributions. In the study and re-study, something uncanny does happen that differs from what has been studied before. Our learning continues every time the name of Freud is archived. Certainly, the purpose of this inquiry is to understand Freud differently than others have, even though he is highly archived and widely discussed still. Jean Laplanche (1999) remarks that—in fact—*all thought is repetitive.*

> Does this mean that all thought is repetitive? Certainly. In the ideal case, let us hope that it is in a state of relative expansion, or at least that it unfolds on planes that, in spite of everything, are changing. (p. 13)

Let us for a moment dwell on Laplanche's claim. All thought is repetitive. It is and yet it is not. Whose Freud? That is the question. And that is the point.

Sigmund Freud felt that he was a stranger in his homeland of Austria and took flight into the intellectual and into the world of psychoanalysis, a world he invented, a world he shared with colleagues and patients. Like Arendt and Mahler, like Kafka, Roth, Zweig, and Schulz, Freud felt alienated. He worked against the grain of culture to carve out intellectual territory in a hostile land. For Freud, the worst hostility he encountered was at the university. Many narratives of European Jews who attended university attest to this fact, as I will show later in this book. Universities are supposed to be places of higher learning. But perhaps they are places where hatred(s) become more highly intellectualized. And it is this that troubles. Universities, unfortunately are places of hate. And this hate creates the dystopia addressed in the last chapter of this book. Hate creates a place of chaos and uncertainty for Jewish scholars. Hate creates a question about intellectual work, and who has the right to do intellectual work, whose hatred gets held in check, whose hatred is allowed to filter into tenure and promotion decisions. Freud (1995) mentions several times both in his autobiography and in his address to the B'nai B'rith that he felt despised. Freud (1995) tells his fellow Jews at the B'nai B'rith,

> I felt as though I were despised and universally shunned. In my loneliness I was seized with a longing to find a circle of picked men of high character who would receive me in a friendly spirit in spite of my temerity. Your society was pointed out to me as the place where such men were to be found. (p. 266)

Freud was despised for being Jewish as well as doing new kinds of intellectual work. The new is never acceptable in places where tradition is upheld over against innovation. Many universities are places where tradition is all and the new means termination. Freud was also despised because he was carving out new and dangerous intellectual material that critiqued Victorian mores. Austrian society historically was notoriously conservative, especially during the reign of the Habsburgs. On the other hand, heterodox minds and radical art movements sprang from Austrian culture as well, especially during the *Fin-De-Siècle.* There is no telling why Austrian culture seemed so contradictory and so extreme. Some of the worst people came from Austria, that is Hitler; while some of the best minds came from Austria as well. Freud reflected the best Austria had to offer. He wrote what he wrote in direct response to conservative Vienna, he wrote what he did against the backdrop of a reactionary culture, and in direct response to his being an ethnic Jew. In a letter to Enrico Morselli, Freud (1960e) explains,

> I feel as though obliged to send you my personal thanks for it. I am not sure that your opinion, which looks upon psychoanalysis as a direct product of the Jewish mind, is correct, but if it is I wouldn't be ashamed. Although I have been alienated from the religion of my forebears for a long time, I have never lost the feeling of solidarity for my people. (p. 365)

Like many Jews who face hostile surroundings and feel consequently isolated and alienated, Freud used these feelings in order to create. He turned negative feelings into productive sources of creativity. Moshe Gresser (1994) tells us that

> Freud conveys to Abraham the conviction that being Jewish in a Gentile world is in a way an advantage, because it is an irreplaceable source of "energy" that helps stimulate one's productivity, bringing out the best capacity for achievement. (p. 137)

Freud speaks to these issues in his address to the B'nai B'rith as he claims that being Jewish allowed him to confront in a uniquely Jewish fashion, in a way that was oppositional to the host culture. Being Jewish allowed Freud to be oppositional and heterodox. A non-Jew would not have written what Freud wrote. Freud (1995) says to his Jewish brothers,

> Because I was a Jew I found myself free from many prejudices which restricted others in the use of their intellect; and as a Jew I was prepared to join the opposition and to do without agreement with the "compact" majority. (p. 267)

If Freud were not Jewish, he would not have written what he wrote, he would not have carved out the discipline that he did. Being marginalized forces one to write from the margins. Where else would one write? Certainly not from the center. And when one writes from the margins, one's topics differ from those coming from the center.

Although Freud addresses his brothers at the B'nai B'rith it is a well-known fact that Freud was not a religious Jew, in fact he detested religion because he argued that it was infantile and superstitious. Yet, on the other hand, Freud did not dismiss out of hand his own Jewishness and seemed proud of his ethnicity. Freud did not convert, as did Gustav Mahler, although there are accounts that tell us that he thought of doing so. Freud's later work was devoted to the study of Moses, interestingly enough. Here again the contradictory nature of Freud's enterprise. Why the interest in Moses if Freud was not a religious Jew? *Moses and monotheism* (1967) is hardly scientific. Freud insisted his work was science. But what does Moses have to do with science? Contradictions are found throughout Freud's corpus. But that doesn't mean we dismiss him or think less of him because he was a complex man. Although Freud's early education included religious instruction, as Moshe Gresser (1994) points out, he seemed to not be able to read much Hebrew. Another Freudian conundrum.

Freud was Jewish and he invented psychoanalysis. Does this mean, as many people have asked in the literature, if that makes psychoanalysis a Jewish science? This question troubles.

No one ever asks, for example, whether nuclear medicine is a "Christian" science. Is there such a thing as an Islamic science? Why then ask the question about Judaism and psychoanalysis? Anti-Semitism is really what is underneath this question. If psychoanalysis means Jewish science then that would be just one more reason to dismiss it. Some of the continuing antipathy toward psychoanalysis stems from anti-Semitism because people think psychoanalysis an "invention of Jews." On an episode of *The sopranos*, Tony's mother makes an anti-Semitic comment about his psychiatrist and states that "all shrinks are Jews who are out to make money." In popular culture this prejudice pervades. Psychoanalysis is still a hotbed, as it were, of controversy. From whence does this controversy spring? Strong reaction against psychoanalysis is, in part, a result of anti-Semitic prejudice. The fact that these negative and hostile feelings still plague the psychoanalytic enterprise disturbs. There is always already more to it than the idea that psychoanalysis is hocus pocus.

The Viennese, who were hostile to Freud and vehemently indifferent or opposed to his work, now memorialize him, just as they do Gustav Mahler. Today the Freud Museum on Bergasse 19 proudly displays Freud's cigars, photographs. A home video narrated by Anna Freud runs continuously as tourists gaze at the empty rooms with only pictures on the walls. What happened to the furniture, one might wonder? Why are the rooms empty? Interestingly enough, that is not something addressed outright in the exhibit. In the home video, tourists see the happy Freud with Anna and company plus his adorable dogs. The video misleads because tourists see the happy Freud. It is as if Austria was a happy home for Jews; it is as if the Viennese embraced him; it is as if he worked side by side with fellow Austrians. But again, like Mahler, the Austrians would have nothing to do with Freud during his lifetime. In fact, they hated him for he was a Jew and many of them thought he was a pervert. Today Freud is Vienna's favorite son. But why the empty rooms? Why the eerie feeling that there is nothing in this so-called museum? All that is left are photographs and cigars. What a bizarre museum. As is well known, Freud finally fled Austria after his daughter Anna was interrogated by the Nazis. Freud died one year after fleeing to London. Freud was in his eighties when he left. Imagine being 80 and being forced to leave one's home. All the furniture was sent to London. That's where it is! The empty rooms in the Freud museum are rather haunting and chilling. These empty rooms signify a hole in the heart of Austria. Mostly all Jewish psychoanalysts fled Austria by 1939. And most of Freud's colleagues were Jews who practiced psychoanalysis. Psychoanalysis was decimated in Austria, nobody of note is there today. As a matter of fact today not many Jews live in Austria and not many Jews who do live in Austria practice psychoanalysis. Why would Jews live in Austria now? If they do live there, what kind of life would that be? The memories are still there. The older generation of ex-Nazis (or Nazis who don't admit their hatred) still roam the streets. What is the younger generation of Jews and non-Jews to think? Some of the younger generation are even more hateful than their parents. Think of the Haider controversy. The David Duke of Vienna lives. He has many followers. Neo-Nazism is alive and well in Central and Eastern Europe. Do not think that it is not.

While Freud lived and worked in Vienna, he experienced much pain due to the hostility he encountered in Viennese society. Yet, as Patrick Mahony (1998) points out, "to various friends he acknowledged that he needed some measure of physical or psychic pain in order to create. His greatest period of creation, in fact, was attended by serious discomfort" (p. 37). Here, Mahony could be talking about Freud's cancer, yet he could also be talking about social discomfort. Freud did not flee, however, into solitude, although he does complain in a letter to Jung that he felt isolated. Robert Coles (1992) writes that

> Sigmund Freud was no solitary doctor or intellectual. He developed intense friendships among his colleagues both before and after his career as the "first psychoanalyst" began, in the late 1890s. His associations with Breuer and Fliess are, by now, the stuff of biographical legend, with Sachs and Jones and Reik: A series of names that have entered history largely by association with Freud. (p. 105)

Intellectual ideas are born of relations. Clearly Freud's relations with others shaped his thinking, as he certainly shaped the thinking of others. Good intellectual work can only be done in-relation to other(s). One cannot do intellectual work alone. One needs intellectual companions, one needs to draw on the work of others, and one needs to cultivate friendships with others in order to talk through ideas and work ideas out. Freud's friends were mostly Jewish. This is an important point to make. It is hard for an American to imagine just how hostile and anti-Semitic Viennese society was and probably still is. So it is no wonder that Freud stuck with his own. Anna Freud (1992, cited in Coles, 1992, p. 18) explains,

> "Yes, the Nazis' anti-Semitism was the worst ever . . . but even there, you have to realize that we had lived our entire lives under the shadow of a really shrill anti-Semitism in Austria. Today, here [in England and America] you would be horrified if you heard even 5 percent of what we had grown accustomed to hearing." (cited in Coles, 1992, p. 18)

Freud responded to the Nazis with anger and sarcasm. One of Freud's sons remarked that the only time he saw his father lose his temper was when he read about the Nazi accession to power in the newspaper. When the Nazis "interviewed" Freud before his exile to England, Moshe Gresser (1994) tells us that Freud responded sarcastically.

> On 4 June 1938 Freud was finally permitted by the Nazis to leave Vienna when they insisted that he sign a statement that he had been well treated by the authorities, he added a sentence after his signature that shows the strength of his defiance, even at the age of 82: "I can recommend the Gestapo most highly to everyone." Martin Freud notes that the style is that of a commercial advertisement, and he reports that its irony was missed by the Nazis as they passed the certificate from man to man, shrugged their shoulders, and marched off. (p. 235)

Freud, unlike his father Jacob, was not passive in the face of aggression. Here, he shows that he confronted his oppressors, albeit in a subtle way. The Nazis were too stupid to understand Freud. Confrontation— of whatever sort— is necessary in the face of hostility. And yet for Jews in Europe it was just too dangerous to confront the Nazis. To confront meant death. On the other hand, there were uprisings (i.e., the Warsaw uprising) and underground movements to try to thwart the Nazis. But the Jews simply did not have enough fire power to confront tanks and machine guns. One cannot fight the enemy with potatoes. Resistance though abounded but ultimately failed because the Jews just couldn't take on such a powerful opposition.

Freud's work was done against a backdrop of increasing hostility toward the Jews. Yet he continued his work. What was it that inspired his work? Ernest Jones (1959), in his well known controversial biography of Freud, comments,

> A notable feature was his preference for comprehensive monographs, on each subject over the condensed accounts given in textbooks, a preference which was also prominent in later years in his archaeological reading. He read widely outside the studies proper, although he mentions that he was thirteen before he read his first novel. (p. 21)

One of the reasons that Freud is a continual lure, is that his wide reading is reflected in his writing. Freud's scope is incredible. The depth in Freud's work draws people in. The more depth and scope, the more discussions abound. Freud, although a specialist in issues psychological, approached his subject as a generalist might. He writes on everything from Moses to jokes. His reading included everything from Goethe to Greek Myth. Peter Gay (1998) comments that Freud had a vast library and money was no object to him when it came to buying books. Gay (1998) tells us,

> This was the poverty-stricken young physician who bought more books than he could afford and who read classic works into the night, deeply moved and no less deeply amused Freud sought out teachers from many centuries: the Greeks, Rabelais, Shakespeare, Cervantes, Moliere, Lessing, Goethe, Schiller, to say nothing of that witty 18th century German amateur of human nature Georg Christoph Lichtenberg, physicist, traveler, and maker of memorable aphorisms. (p. 45)

Freud engages in what William Pinar (1999) terms "open intellectual exploration" (p. xvi). Like psychoanalysis, curriculum theorizing requires scholars to engage in open intellectual exploration and draws on intellectual work from many different disciplines. This is what makes curriculum studies so difficult and yet exciting. But generalists who draw from interdisciplinary texts tend to get critiqued by conservative colleagues in the field of education because our interests are not narrowly focused on the schools. Of course our interests and concerns are of the schools, but in order to speak of school life one must look to the larger culture in order to better understand school culture. As William Pinar (1999) forcefully argues, we are not professors of schooling, but rather we are professors of education. Studying education, for curriculum theorists, means studying the larger culture in which education is carried out. That means studying people like Freud!

One of the interesting things about Freud is that he saw the connection between what it is that educators and psychoanalysts do. Raymond Dyer (1983) tells us that,

> in a letter to Pfister (Meng and E.L. Freud 1963, letter dated January 1, 1913), [Sigmund] Freud insisted on defending "the rights of educationalists to analysis" (p. 59), and a month later he contributed the introduction (1913b) to a book by Pfister, in which he noted that education and therapy had a definite relationship. (p. 18)

The links between education and therapy are metaphoric not literal. Teachers do not literally do therapy with students. However, a psychoanalytic frame of mind may shape interactions with students on a profound level. Freud should be integrated in every school curriculum. What better way to understand self-formation than by studying Freud. But as we know, American curriculum is anti-intellectual and Freud is studied little. Academic psychologists are hardly interested in Freud. And this is the shame that is the American Academy.

Psychoanalysis and Curriculum Studies

Psychoanalytic thinking is a certain way of positioning oneself in the world and in the classroom. Psychoanalytic thinking is a way of being and seeing. What one learns from psychoanalysis is that thinking historically, genealogically, archaeologically, and culturally becomes necessary when dealing with issues in the classroom. Curriculum as historical and relational text, curriculum as cultural, archaeological, and genealogical text shapes classroom discussions in ways that are deeper psychologically and philosophically. Thinking psychoanalytically is profoundly object-relational. The misguided assumption that psychoanalysis is primarily solipsistic is simple-minded. Thinking psychoanalytically means thinking in terms of relation-to-the-other. The ways in which one thinks about the other can alter the ways in which one thinks about the self. Christopher Bollas (1997) speaks to this point.

> I tried to get to that side of the equation of object relations by conceptualizing the other as a process, as a transformational object. I do have a project to try to delineate and differentiate, among the internal objects we hold in us, those that are fundamentally a result of our own work, and those that might be more fundamentally the work of the other upon the self. (p. 57)

Naomi Rucker (1998) teaches that subjects work *through* each other, not *on* each other. Students work *through* teachers as teachers work *through* students psychologically. Subjects work through other subjects, as Rucker (1998) might say. As against Rucker, I prefer the traditional language of object-relations whereby subjects work through objects, because this notion of the object helps distinguish between subject and object, self and other. Further, I do not know whether the other always becomes a subject for me. I do think that the other is always already an object to me because in the psychological sense, I cannot fully understand the other as subject, I can only understand the other as other, as not-self. At any rate, the work of the other upon the self and the self upon the other in the classroom becomes intensified when uncomfortable issues are raised. When students have their taken-for-granted assumptions ruffled, it becomes difficult to say just how those internal objects are transformed. When professors are shaken by students' acting out in the context of difficult material, teaching can become psychically damaging if the teacher is not prepared to deal with negative transference. At any rate, these transformations between students and teachers happen over long periods of time and hopefully the transformations will be positive ones, but one can never underestimate the strength of negative transference. Perhaps the educational experience is inevitably filled with conflict, especially if professors and teachers push students to undo their prejudices and look into their hearts of darkness.

There is something about the teaching enterprise that is connected with the notion of taking time, or letting time be so that the process of becoming educated is a process that is fully experienced. This is not dissimilar to the experience of undergoing analysis. Michael Eigen

(1997) suggests that we

> relax into giving time a chance to develop its own flow, and allow that there should be a "later" . . . or a "then," or enough room or enough of a gap, so that some kind of approach of movement from here to there is possible. (p. 127)

Eigen's remarks are interesting when applied to the space of the classroom. Time and education are intricately tied. It is misleading for students to believe that education is like fast food. Rather, education, original thought, creativity, and invention take time, take years. The paper written in twelve weeks does not have time to take on its own life, to really expand into the unfolding of being. Students get the wrong impression when they believe that this is how scholarship is done. Good scholarship takes time and movement and one must relax into it and let it be, as Eigen says. One must not rush thinking. Contemplation and dwelling need time, need space, need nurturing and prodding. Some ideas do indeed come later, in the then-time. Some ideas never come at all. But one must make room for the ideas that do come and tend to them. "And then the idea came to me that.". . . It is unfortunate that students have to be rushed and that teachers must rush through material so as to meet deadlines. The rush to meet deadlines creates gaps in our educative experience. There is only so much one can do in a semester or a quarter. However, these gaps can be both positive and negative. Negative gaps are ones that perhaps may never be filled. At some point, one simply cannot make up for lost time. But still, one must not think that one can entirely fill gaps either. One works through gaps and works with them as part and parcel of one's thinking. Intellectual work creates gaps by its very nature. Focusing on certain issues creates gaps whereby other issues get excluded and edited out of the conversation. But this conversation can be transformational. Transformational intellectual work is work that changes one and transforms others by making others think differently.

Transformational work that happens over time is by no means solitary. The work of the intellect happens in relation between self, the other, and within institutions. One's subjectivity, in other words, is always built upon these premises. Subjectivity does not mean solipsism. Felix Guattari (1995) suggests that "the emphasis on subjectivity [is] . . . the product of individuals, groups and institutions" (p. 1). The self is always already a social self. And the social self is always already one injected with internal objects, transformational objects, and psychic slippage between one object and another, between one self and another person. The enterprise of education complicates subjectivities because it adds a third element to subject building: Books. Our lives as scholars are ultimately about our lives with texts. One engages with texts as one engages with the other. The writer of the text and the reader of the text are complicated creatures who dwell in the doubled space of internal and external object relatedness. Norman Finkelstein (1996), in an interesting piece titled "The Master of Turning: Walter Benjamin, Gershom Scholem, Harold Bloom, and the writing of a Jewish life," contends,

> Lives are like literary works: we live as we read, experiencing the life and the text with an ineluctable doubleness. The life and the text stretch before and behind; the result, if not narrative, allowing forward and backward movement over time. But the life and the text also present themselves as momentary: we encounter an uncanny simultaneity of events. (p. 415)

The simultaneity of events—psychically—is that of living in dual worlds, the world of the everyday and the world of textual encounters. Psychologically this experience of dual worlds is felt in uncanny ways. Sometimes one reads so much of a writer that parts of the writer get injected, perhaps unconsciously, into the self. Naomi Rucker and Karen Lombardi (1998) explain this process as it occurs between two people. "Within an object-relational paradigm, concordance refers to a process of introjection, through which parts of the other are experienced as existing within the self" (p. 29). This fusion or con-fusion between reader and text is common. Although, one may think in Freudian terms, one must always be vigilant not to "become" Freud, as it were. That is, when one does a Freudian analysis, it is important not to merely repeat what Freud has said— to be a slave to mimesis— but rather, to make Freud's texts one's own. But the problem of being devoured or swallowed up by the Freudian corpus is always there. When one is too close to Freud's texts, when one is "in concordance" trouble ensues. Being too close to the text forces one to lose one's sense of self. Being too close to the text gets reflected in mimetic scholarship. If one only engages in mimetic scholarship whereby Freud says x and I say that Freud says x, Freud says y and I say that Freud says y, generativity gets thwarted. Strong identifications with literary figures pose these sorts of exegetical dilemmas. Jane Gallop, at a session at the Modern Language Association (2003 held in NewYork City), spoke of falling in love with her subject. Positive transference (or in this case falling in love with Roland Barthes, as Gallop put it) is a result of concordance and introjection. Falling in love means that parts of the other (psychologically) do indeed move through the self. Falling in love with Freud is not a bad thing, if one keeps one's sense of self intact and is not devoured by Freud. One mustn't lose one's head over Freud, as it were. Freud was well aware of these potential pitfalls and hated the genre of biography for these reasons. Freud did not want to be idolized. And Sophie Freud, his granddaughter, pointed this problem out to me (personal communication). Yet the biographer tends to idealize her subject. In a well-known letter to Arnold Zweig, Freud (1960f) states,

> Anyone turning to biography commits himself to lies, to concealment, to hypocrisy, to flattery, and even to hiding his own lack of understanding, for biographical truth is not to be had, and even if it were it couldn't be used. (p. 431)

Perhaps there is no biographical truth with a capital T. Rather, biographies or life histories allow one to raise questions about someone else's life in order to raise questions about one's own. One learns about the process of studying the self through the study of the other. Scholars get uncanny clues to self-understanding while digging through the archives— metaphorically speaking— that reveal yet conceal. Sometimes, one discovers what has not been understood before. Other times, scholars are not so lucky. Hours of searching in the dusty stacks of the archives can be frustrating because one never knows where the path of research will lead or what dead ends will block the way.

Education as an archival text, is an active engagement with digging—digging through one's multifaceted registers of self-understanding—ironically *through* studying the lives of others. The passion for studying, the passion for engaging in theoretical engagement with texts is an educational commitment, a life commitment. Theoretical engagement helps scholars and students alike to live more full lives, to live richer, more productive lives. To allow students to engage freely in phantasy-life is part of the process of digging through the archives of self. To allow students to engage freely in speaking associatively and dreaming dreams is crucial to growth. Not all psychoanalytic texts forefront phantasy. Klein, though, did forefront phantasy and suggested that an infant's phantasy life coincides with birth. Thus, Kleinian texts would be a good place to start to grapple with the intersections between education, phantasy, and life history.

Students are so used to not thinking about themselves and dismissing the importance of autobiography and phantasy. Studying Kleinian texts with students might be a breath of fresh air!!

Educators might begin thinking about what it is we use theoretical constructs to do. If students are taught to dream and to utilize phantasy to discover who they are, this might enhance their creative abilities. What educators might avoid, though, is teaching metanarratives that are unrealistic or idealistic or that miseducate students about the way the world is. Adam Phillips (2000) says,

> Indeed, it is part of the moral gist of their work [Darwin and Freud] not merely that we use our ideals to deny, to over-protect ourselves from, reality; but that these ideals—of redemption, of cure, of progress, of absolute knowledge, of pure goodness,—are refuges that stop us living in the world as it is and finding out what it is like, and therefore what we could be like in it. (p. 17)

Our world is not a world of redemption or progress. Our world is one of wars and cruelty, one of great happiness—yes—but also of grave sadness, of illness and disease, of death and destruction. To find out about the grave misfortunes of the underclass, to learn about the struggles and torments of subaltern peoples is to find out that the world is cruel. There are no places of refuge in the world because it is a violent, crazy place. We live in a generation that has witnessed 9-11, the war with Iraq, nuclear proliferation. Undergraduates have witnessed more terror than did my generation growing up. The world seems to be spinning out of control. Current events have taken on sinister and cynical dimensions rather quickly. The world has a death instinct; the world is the death instinct writ large. We are out to destroy each other and the planet.

In the midst of the turmoil of a post–9-11 world, educators push onward. To teach young people the love of reading and the love of studying becomes paramount in such confusing times. We may not be able to teach students how to develop their own ideas, but we can teach them the value of reading and the value of deep and structured studying. We can teach them the dangers of colonization in their own thinking. Teachers can teach students to not allow academic institutions to colonize their personhood. Academic institutions should exist to allow young people to develop their own thoughts in their own time, not colonize them as subordinate subjects of learning. Unfortunately for Jewish students, things are made more difficult because of ongoing battles with anti-Semitism. This was true in Freud's day as it is in our own. What we learn from Freud is that because of his experiences with anti-Semites he was forced to develop a rich phantasy life and ultimately a new discipline in order to survive a hostile environment. When Freud was a student, he began to develop his own ideas and thinking partly because of and perhaps despite the opposition he felt because he was Jewish. Peter Gay (1998) tells us,

> In the upper classes of his Gymnasium, Freud, too, began to recognize "the consequences of being descended from an alien race." As the "anti-Semitic agitation among school comrades admonished me to take a position," he identified all the more closely with that hero of his youth, the semite, Hannibal. (p. 20)

Freud fantasized himself a warrior, a conqueror. Freud teaches the value of analytic thinking. He teaches much about living as a Jewish intellectual in the midst of turmoil and opposition, chaos and uncertainty. Freud teaches the value of stick-to-it-ness, despite living in a hostile culture. It was not easy for Freud to make it, because the Viennese treated him badly. He was not recognized by Austrians. The pain of not being recognized is intellectually devastating and emotionally ruinous. But somehow Freud kept on working. Take courage from Freud. Keep working, even if surrounded by hostile colleagues in a hostile institution. Keep writing. Be enraged. Feel anguish. But make use of it all through scholarship and teaching.

Being Jewish means being misunderstood about Jewish life. Jews live differently than Christians in American culture. Jews have to maneuver as minorities within a majority Christian culture. And it is this that many Christians do not understand. Jews live in two realms at once. Being part of a subaltern group, Jews must develop a parallel existence. One lives on two separate registers at once. This parallel existence is what Jonathan and Daniel Boyarin (2002) term "diasporic consciousness."

> This is the paradoxical power of diaspora. On the one hand, everything that defines us is compounded of all the questions of our ancestors. On the other hand, everything is permanently at risk. Thus contingency and genealogy are the two central components of diasporic consciousness. (p. 4)

Diasporic consciousness, the consciousness of exile demands reflection around one's exiled state. One's exile and displacement beckons study. To live in the now as a subaltern people is to realize how fragile we are. When one lives on the edges of experience as the Other, one becomes more cognizant of the fragility of life. Subaltern peoples are always already at risk because of being targets of hate. Thinking historically and culturally empowers one to move forward through life, not to sink in the chaos of hate. Studying, in a sense, frees one from oppressive culture. Studying allows you to imagine otherwise.

Curricular questions are always already questions about how we come to be who we are by studying ancestors. Knowldeges of subaltern peoples is changeable and open to contestation. Ours is an ever-changing labyrinth. Jews are always already at risk. We are at risk of being misunderstood. Some might think our work too narrow or of only Jewish interest when clearly it is not. If anything, work on Jewish identity concerns that identity in-relation to non-Jewish peers, students, colleagues, and teachers. We must live in-relation and attempt to understand our place in the world, even and especially when it is a hostile one. Students and teachers alike must learn to live in relation with the other in that place we call school or university.

Melanie Klein and Martin Buber: Phantasy and Genuine Meeting

Jewish intellectuals create what they do because of who they are. Jewish intellectuals create what they do because of the historical context in which they are thrown. Melanie Klein, like the other Jewish intellectuals I have cited, felt estranged because of her Austrian Jewish background yet she claims that it was this that enabled her to write as boldly as she did. Klein (2001, cited in Kristeva, 2001, pp. 22–23) remarks that

> I was glad to confirm my own Jewish origin, though I am afraid that I have no religious beliefs whatever. . . . Who knows! This might have given me the strength always to be a minority about my scientific work and not to mind, and to be quite willing to stand up against a majority for I which I had some contempt. (2001, cited in Kristeva, 2001, pp. 22–23)

Klein, the mother of object-relations theory, had to take bold steps to do her own version of psychoanalysis in the face of psychoanalytic orthodoxy, as well as in the face of anti-Semitism. Like Sigmund Freud, Klein loathed the masses, loathed mediocrity and conformity. Like Sigmund Freud, Melanie Klein was not religious, but felt wedded, so to speak, to Jewish culture and Jewish ethnicity, although she eventually converted to Christianity. Perhaps psychoanalysis was a sublimated form of religion for many European Jewish intellectuals. Phyllis Grosskurth (1986) argues, in the case of Klein,

> While always feeling "Jewish," she was never a Zionist, and her way of life was in no way distinguishable from that of a Gentile. Yet as a Jewish child in Catholic Vienna she must have been acutely conscious that she was an outsider and a member of an often persecuted minority. Psychoanalysis became for many Jews a religion. (p. 14)

Not only was Klein an outsider in Viennese culture as a Jew, she was an outsider within the psychoanalytic movement because her ideas were heterodox and bizarre even for other analysts. Klein's notions of phantasy, the early arrival of the Oedipus Complex, greed, and the primacy of the mother over against that of the father-phallus, and the necessity, as Kristeva (2001) points out, of matricide (figuratively, that is) and envy brought her much scorn. Meira Likierman (2001) tells us that

> when Klein first presented her thinking on envy to her colleagues, Winnicott is reported to have uttered despairingly, "oh no, she can't do this!!" One of her most staunch adherents, Paula Heimann, already increasingly estranged from Klein, would come to regard this moment as making an irrevocable break between the two of them . . . these reactions aptly indicate the misgivings with which the concept was generally received. (p. 172)

Klein was clearly a maverick. There is no one like Klein. One might continually come back to her because one does not understand her. There is such incredible richness in her writing. Her phrases ooze with nuance and depth. Disturbingly, Klein's work is hard to find in bookstores, even in cities where many analysts work. Perhaps she is still too shocking! In a piece titled "Some theoretical conclusions regarding the emotional life of the infant" (1952/1993), Klein boldly states thus:

> In his destructive phantasies, he bites and tears up the breast, devours it, annihilates it; and he feels that the breast will attack him in the same way. As urethral—and anal—sadistic impulses gain in strength, the infant in his mind attacks the breast with poisonous urine and explosive feces, and therefore expects it to be poisonous and explosive towards him. (p. 63)

For American readers who are used to sanitized curricula, this citation has got to shock. Klein sanitized nothing. She said what she thought and did not care about what others thought. What interests about her thinking is her boldness and her audacity. Are infants this destructive? Can infants phantasize in the Kleinian sense? Do they? How to tell? If phantasy comes before language how can one show that infants phantasize? They cannot tell you in words that they want to devour mothers' breast. But perhaps this is not the important issue at hand. What is important is that Klein thinks like no one else. She makes one think differently. What she attempts to communicate is striking. What one might take from Klein is her conviction that human beings are not so nice, in fact, many are sadistic. If anything, Klein teaches us that children are not "innocent." Children are subject to the same unconscious stirrings as are adults. "Subject" to the unconscious is key here. We are in the shadow of the object and that shadow is the driving force behind phantasy, thought, imagination, feeling. Children are subject to the death drive just as are adults. Children express themselves in complex phantasy life. Children's play is loaded with symbolism. When children rip up their dolls or play war with *G.I.Joe®* it all has symbolic meaning. A child who rips up her doll may be annihilating her mother! It is astonishing that Klein's work was virtually a forbidden topic in America in the 1950s and 1960s. Balint and Winnicott seem relatively tame compared to Klein. Both of these theorists are more palatable than Klein. I prefer Klein to these men for several reasons. I find her interminably interesting and insightful, biting and unrelentingly shocking. Neither Balint nor Winnicott have the emotional effect on me that Klein has, although I think their work immensely important.

What is also interesting in the context of this study is that Klein was influenced intellectually by her father, much like Buber—who was influenced intellectually by his mother and grandmother. Klein's father was radical for his time. He made a radical shift in his life from being an orthodox Jew to becoming a medical doctor. From the ultra religious to the ultra scientific! Likierman (2001) tells us that like her father, "Klein was to become intellectually rebellious and balk at narrow traditionalism all her life, in her case, the dogmas of the psychoanalytic establishment" (p. 28). Klein certainly was rebellious and original like no other. Maybe the fact that she did not attend university *helped* her. She was not bombarded with canons of orthodoxy in conservative institutions which often thwart creative potential. Neither Melanie Klein nor Anna Freud graduated from university for that matter, and yet they created incredible work. One simply cannot imagine the obstacles women had to face in Eastern and Central Europe to gain any sort of recognition, especially if they were not medical doctors and not university trained. It is a testament to both of these mothers of psychoanalysis that they not only got recognition but founded two different schools of psychoanalysis, Anna Freud is often assumed the mother of ego psychology, while Melanie Klein is considered the mother of object-relations.

The Jewish intellectual who examines Jewish life in the context of education per se is Martin Buber. Although Martin Buber was a philosopher, he was also interested in psychoanalytic ideas as they related to social philosophy and education. Judith Buber Agassi (1999) explains,

> Buber's lifelong interest in psychology, psychopathology, and psychotherapy was tied to his philosophical concerns, religious, ethical, and anthropological. He was neither a psychotherapist nor a psychologist. As a young student he had studied three semesters of clinical psychiatry in Leipzig and Berlin and with Bleuler in Zurich, yet he never seriously considered a professional career in psychiatry. (p. vii)

Buber's social philosophy is deeply psychological and he is devoted to fleshing these ideas out when talking of education. Buber teaches how to think about education in more psychodynamic and philosophic terms. My intellectual interests run close to Buber's in that I have been steeped in the Western philosophical tradition as well as in the psychoanalytic one. But my primary concerns, like Buber's, are educative. My primary concerns turn on teaching the young, and—in a sense—caring for the young. I work to inspire them to think more deeply about what it means to be an educated person and what it means to become a teacher. Public school teachers are not often thought of as intellectuals, as has been well documented. But I would like to get them to think that they can be intellectuals and can work creatively to shape their own students' lives.

Although Buber could not be more different from Freud, he did have some interaction with him. In fact, Paul Roazen (1999) comments that

> both Freud and Buber were Austrians and Jews; it would be impossible to understand either of them apart from their cultural and religious backgrounds. Out of the remarkable cosmopolitan maelstrom of old world civilization they went in startlingly opposite directions. (pp. xx–xxi)

These thinkers had very different concerns of course. Freud was hostile to religion, yet tied to his Jewish roots. Freud was an ethnic Jew wedded to Enlightenment ideals. Buber, on the other hand, was a religious thinker who was neither assimilated nor wedded to Enlightenment thinking. Leon Botstein (2002) remarks,

> The premodern shtetl Jew emerged as the object of idealization, an ironic reversal of the contempt German Jewry had for traditional Eastern European Jews. In lieu of the ideal of the Jew as good European and cosmopolitan world citizen, German speaking Jewish writers of the 1920s as diverse as Martin Buber, Alfred Doblin, Arnold Zweig, and Joseph Roth held up the traditional Ostjude, steeped in religion and proud to stand apart from the Gentile world. (p. 6)

In many ways Buber was premodern. He was steeped in the Hassidic tradition. Thus, reading Buber is a totally different experience from reading Freud. Buber's writing seems mystical, while Freud's work seems literary as well as scientific. Buber's writing seems theological, while Freud's is theoretical. Buber's writing is grounded in the everyday, while Freud's is grounded in the abstract and universal. At the same time, Freud's case histories are grounded in the particular, while Buber makes sweeping generalizations about the human condition. While Freud makes few references to education, Buber makes many.

Both Freud and Buber, though, attended the Gymnasium. Both were schooled in similar kinds of educational institutions. Buber's ideas about curriculum sprang from his own educative experience. Buber (1948/1976) states,

> Today what was once matter of course—our language, the Scriptures, our history—must become curriculum of the most crucial importance. The passion to hand down can be replaced only by the passion to study . . . and thus re-establish the bond of memory that joins the community together. (p. 148)

The passion to hand down is the passion of memory. Memory and history are intertwined but not the same. History is guided by strict rules, memory has no rules. Memory is a concept that is increasingly important in Jewish scholarship (Morris, 2001). Memory is linked to history, as memory is the oral recounting of a peoples. History is the written, systematic approach to memory. Since the Jewish people have been a people of the Diaspora, writing historical texts has been nearly impossible prior to the twentieth century. Jews had been exiled from almost every European country through pogroms and persecution and were forcibly scattered here and there across Europe. The quest for memory, therefore, is an attempt to capture the stories of the

Jewish people by way of oral tradition. Memory work, then, becomes a sacred duty. Therefore, to study ancestors and genealogy is an ethical calling. Without understanding where we have been as Jews we do not know where we are going. Unfortunately, the passion to study and the passion to learn gets thwarted in the American Academy because of the increasing corporatization of the university. Teachers are so busy fulfilling service requirements and worrying about NCATE and SACS that they tend to forget about why they became teachers to begin with. How can teachers inspire young people when they are saddled with Kafkaesque bureaucracy? An intellectual atmosphere is rarely found at the university. Petty bureaucrats abound. Kafka's character K was right after all! Can petty bureaucrats teach at all? The university likes teachers who are technocrats. The university of technocracy is not a place of higher learning. It might be a place of lower learning. How uninspiring. If students cannot be inspired by teachers who will they become? They will be inspired to become technocrats and nothing more. Many students reflect the larger anti-intellectual American culture. They are apathetic, they have little passion for study, and they hate to read. How can we teach people like that? Can we teach anything? In a well-known piece by Franz Rosenzweig (1955) titled "Toward a Renaissance of Jewish Learning" he states thus:

> We have no teachers because we have no teaching profession; we have no teaching profession because we have no scholars; we have no scholars because we have no learning. Teaching and study have both deteriorated. And they have done so because we lack that which gives animation to both science and education— life itself. (p. 60)

Rosenzweig's claim is unfortunately right on the mark even today. Particularly in the field of education, anti-intellectualism abounds. Teachers are not public intellectuals, although we would like them to be. Teachers are more like testers. Teachers are clerks who administer exams. Teachers are like workers at the driver's license bureaus. Teachers are not educating students—they are teaching them to take prefabricated standardized exams. This is not teaching, this is a travesty. The state of public education today is a catastrophe! American educators must work continually to undo the madness of standardization. Yes, teaching and studying have deteriorated. Preparing for standardized tests is not study!! The world is NOT a standardized test. Should not students have a clue to what the world is about in the twenty-first century? Should not they be studying cultural issues? If they are not, where will the next generation be? What will they be able to do? What kind of thoughts will fuel their deeds? Study entails deep introspection and reading. Serious reading requires time and energy. Thinking can only happen when one can think freely across texts. Study means grappling carefully with texts, with the words of others, with meaningful words. One comes to one's insights only by way of carefully crafted thoughts that take time and care. Caring for one's thoughts is like caring for one's children. There are no standardized tests for raising children. Or are there? Schools are dead places when the goal is raising test scores, not raising original thinkers. Universities are madhouses when professors are forced to align their syllabus with the State. One only needs to think of Nazi Germany to make the connections.

At any rate, Martin Buber inspires. He loved study, he loved words, he loved books, he loved thoughts. His words are inspiringly eloquent. Perhaps if teachers read Buber they too will be inspired to once again find the passion to study and read and pass these practices onto their students. But education must be re-injected with life for anyone to find passion there. Buber (1948/1976) speaks with passion about educating the young. And he says that teachers' responsibility is tremendous. It is tremendous because the task at hand is to allow students to "unfold" (p. 77) while they are in our "care." Buber (1948/1976) says,

> The educator whom I have in mind lives in a world of individuals, a certain number of whom are always at any one time committed to his care. He sees each of these individuals as in a position to become a unique, single person, and thus the bearer of a special task of existence which can be fulfilled through him and him alone. (p. 76)

The relation between student and teacher, says Buber, is one of "genuine meeting" and surprise. Buber warns against prepackaged thinking, thinking that is not spontaneous, thinking that is not authentically of the moment. "Real meeting" is the meeting between individuals where spontaneous conversations are allowed to arise. So often, schooling mitigates against real meeting. Buber says, "Dialogue [or what he also calls genuine meeting] in my sense implies of necessity the unforeseen, and its basic element is surprise" (1948/1976, p. 190). Surprise cannot come in a package of lecture notes; surprise happens only in open-ended dialogue with students. Allowing students the full range of interpretability is the key to surprise. We must never think for our students, they must think for themselves.

We must never interpret for our students, they must learn to interpret for themselves. What is genuinely surprising about teaching is understanding the Otherness of the other, the Otherness of our students and the Otherness of their interpretations of texts. One of the major themes in Buber's work concerns the importance of meeting the other and embracing one's Otherness. Buber (1992) states,

> Only when the individual knows the other in all his otherness as himself, as man, and from there breaks through to the other, has he broken through his solitude in a strict and transforming meeting. (p. 38)

The point Buber makes about not retreating into solitude is crucial. Buber's thinking is profoundly social. The self only knows the other in genuine meeting with the other's Otherness. Education is about understanding this. Curriculum scholarship is no stranger to grappling with the notion of Otherness. When Buber speaks of the Otherness of the other, he speaks as a Jew about the Jew. For Buber, the ultimate goal of education is allowing the student to understand how she can become self-educated, how she can begin to develop a passion for study, a passion for the book. Buber (1947/2002) states,

> So the responsibility for this realm of life allotted and entrusted to him, the constant responsibility for this living soul, points him to that which seems impossible and yet is somehow granted to us—self-education. (p. 120)

This self-education, as Buber calls it, should be infused with continual genuine meetings and the ever-present address to the other. How do teachers teach students to learn how to become self-educated and thus be able to address the other as other? What does it mean to self-educate? Does schooling allow for self-educating? Perhaps not. And perhaps this is the problem of the American system of education. It does not allow students to even know how to begin to become self-educated. Rather, students are colonized by a tyrannical standardization, by a standardized way of knowing. This has nothing to do with becoming an educated person. In fact, standardization of any kind miseducates.

Buber (1992b) says, "living means being addressed" (p. 49). But it is not clear what this address suggests. When one attempts to understand an address one is always already uncertain as to the meaning of the other's Otherness. Buber (1992b) explains,

> What occurs to me says something to me, but what it says to me cannot be revealed by any esoteric information; for it has never been said

> before nor is it composed of sounds that have been said. It can neither be interpreted nor translated, I can have it neither explained nor displayed; it is not a what at all, it is said into my very life; it is no experience that can be remembered independently of the situation, it remains the address of that moment and cannot be isolated, it remains the question of a question and will have its answer. (p. 51)

In the genuine encounter between student and teacher the address is that which cannot ultimately be translated. We remain forever strangers across difference to one another. But this does not mean that we do not engage in genuine communication, for sometimes we do. Yet, genuine communication does not necessarily mean clarity. Genuine communication is difficult to follow and difficult to understand. Real communication between people reveals and conceals simultaneously; real communication is sometimes no communication. One spontaneous gesture follows another, but do we really know what it is we say to one another?

Unlike genuine communication and genuine meeting, Buber suggests that there are instances of what he terms "mismeeting." Mismeeting is an experience whereby the mother does not love her child. Mismeeting is the experience whereby the teacher does not hear her student, does not listen to the student, does not pay attention to the student. When the teacher does not care for her student she engages in mismeeting. Communication between student and teacher is made difficult because of what Buber calls the "narrow ridge." We walk along the "narrow ridge" nearly falling with every step. The path is rocky and steep. There will be many times we are addressed and do not hear the call; there will be many instances of "mismeeting." Buber (1992a) states, "[o]n the far side of the subjective, on this side of the objective, on the narrow ridge, where I and Thou meet, there is the realm of the 'between' " (p. 40). The narrow ridge between the I and the Thou is that place where mismeeting may occur. We must find a way to get over the narrow ridge, not fall into the abyss. Maurice Friedman (1991) explains that for Buber,

> the "narrow ridge" is a metaphor for human existence itself: an existence in which one must walk with faltering step, threatened at every moment by the danger of falling into the abysses to the left and to the right. (pp. 43–44)

Teaching is a troubled profession because it is filled with many "narrow ridges" both in public education as well as inside the academy. These institutions somehow manage to thwart the educative process.

The institutions of learning are not really about learning, they are about something else. Institutions are like puppy mills where they spit out puppies that are born too soon and weaned too early, as Otto Rank (1996) might have thought. Puppies taken away from their mothers too soon are mis-educated and find themselves in a host of mismeetings because they cannot socially connect with their worlds. Students in education mills are mis-educated because they don't have time to grow into themselves and be inspired by intellectuals who are passionate about studying. Thus, for Buber, our responsibility is to try to get students to engage in real communication, genuine meeting, and to be able to become self-educated. This requires educators to take the notion of responsibility seriously. Buber comments on the notion of responsibility. He states,

> The idea of responsibility is to be brought back from the province of specialized ethics, of an "ought" that swings free in the air, into that of lived life. Genuine responsibility exists only where there is real responding. (p. 54)

There *can* be real responding inside institutions of learning if we do not continually fall off the narrow ridge into the abyss of bureaucracy, into the abyss of service work, into the abyss of standardization. Responding for Buber implies responsibility. Yet, the notion of responsibility is too often thought of in superficial terms. Responsibility for the care of the other, for the nurturing of our young people, for the future of our children, requires an understanding of culture, historicity, identity, nation, globalization, place, memory, race, class, and gender. It is our responsibility as teachers to understand students and their cultures as well as our own culture. It is our responsibility to understand our own identity as well as understanding the identities of others. There is no better way to take responsibility for our children's education than through teaching. Buber (1948/1976) claims that teaching is a social responsibility as we pass on knowledge from one generation to the next.

> The life of the spirit of a people is renewed whenever a teaching generation transmits it to a learning generation which, in turn, growing into teachers, transmits the spirit through the lips of new teachers to the ears of new pupils. (p. 137)

Teaching is NOT strategy and technique. Teaching has little to do with methodology, as curriculum theorists are well aware. Teaching has to do with the transmission of culture to young people. Teaching has to do with inspiring young people *to be inspired* by learning and by reading. Teaching means inspiring one to attain genuine thought and intellectual development. It has long been documented that public schools do not do this. In fact, public schools destroy everything about which Buber speaks. Public schools have fallen into the abyss of standardization and routinization. In light of this fact, Philip Wexler (1996) has called for the resacralization of education. And it is Buber who can help us to "resacralize" education. Buber suggests that "What matters is that time and time again an older generation . . . comes to a younger generation with the desire to teach, waken, and shape it" (p. 139). The coming together of the older generation with the younger one—to engage in genuine educational experience—is the meeting of the I with the Thou in that space of the between. What transpires between the space of the I and the Thou is sacred. Moreover, Buber (1958/1986) suggests that only in this relation between I and Thou can a decision be made. "Only he who knows relation and knows about the presence of the Thou is capable of decision. He who decides is free, for he has approached the Face" (p. 51). The decision to meet one another in an I-Thou relation enables one to make the decision to be a good teacher, to be a good scholar. The decision to meet one another in one's Otherness in the space of the between allows one to see the Face. The Face is the Face of the Other. Of course, for Buber the ultimate Other is the Face of God.

The Face of the Other, when that face is Jewish, is in exile. The Jewish condition is always already one of exile. Buber (1948/1976) remarks,

> Whether or not it is aware of it, this people is always living on ground that may at any moment give way beneath its feet. . . . It is this inescapable state of insecurity which we have in mind when we designate the Jewish Diaspora as galut, i.e. as exile. (p. 167)

Against the backdrop of perpetual states of exile, whether metaphoric or literal, Buber's discussion of I and Thou and of the narrow ridge comes into focus within a Jewish perspective. Genuine meeting with The Thou of the Other is always uncertain because of the condition of exile and the slippery ground upon which we walk as Jews in a hostile world. The fragility of existence heightens when one becomes aware of living on the narrow ridge. This is the existential condition particular to Jewish life. Education for Jews means learning to live in hostile lands while continuing the struggle to be productive, continuing the struggle to become educated, to become learned.

Education is always already a social encounter between people. American educators can learn from Buber how to inject life back into the curriculum. Schools are places that can be deadening. What makes schools come alive is strangeness. Buber refers to students' Otherness as strangeness. This is what makes education come alive. It is this Otherness that teachers must embrace. Life is found in the state of Otherness, not in the state of sameness or complacency.

Buber's work refreshes. The complexity of his thought tends to be overlooked or misunderstood. He is by no means easy to comprehend. Buber offers profound insights into the human condition. Interestingly enough, the Nazis were particularly threatened by Buber and called him an "arch-Jew." Maurice Friedman (1991) explains: "The Nazis called Buber the 'arch-Jew,' a designation that he quoted with pride in his address on the occasion of his receiving the Peace Prize of the German Book Trade in 1953" (p. 215). Sander Gilman (1995b) tells us that the derogatory term "arch" was used frequently in many different ways by anti-Semites. Gilman (1995b) says,

> At the turn of the century, Jews are both the arch-bankers and the arch-revolutionaires, both the false nobility of Paris and the wandering Eastern Jews of Warsaw, all things to all groups who need to define outsiders. (p. 59)

Webster's Dictionary says that the term "arch" has two different meanings. One is a person who is a "chief" or who is of some importance. But the more interesting meaning of arch designates someone who is "cunning, sly; roguish; mischievous" (p. 77). The most Jewish Jew, the arch-Jew (Buber, in this case) in other words, fits all these stereotypes, according to anti-Semites. The arch-Jew is slippery, not to be trusted. There is something sinister about this arch-Jew, according to anti-Semitic lure. Is there such a thing as an "arch-Christian?" I do not think so. Why is the Jew an arch Jew? Why is the religious Jew an arch-Jew? Buber was a religious Jew, unlike Freud. Was Freud too considered an arch-Jew? Some might say yes, even though he was not religious. Why are Jewish bankers arch-Jews? Why are the Rothschilds arch-Jews? Why are Polish Jews arch Jews? I ask these questions in order to amplify the absurdity of the claim that Jews have the quality of being arch. I really do not know what this means and I do not know why this term stuck to Jews and not Christians. However, I do know that even today, it is not acceptable to be too Jewish in Christian culture. Being too Jewish is considered being arch I suppose. Being too

Jewish is giving too much away. Being too Jewish means not assimilating to Christian America. Being too Jewish means being too visible. Maybe just being a Jew is too Jewish. As Svi Shaprio's (1999) work suggests, Jews are strangers dwelling in strange lands. Jews are strangers in strange institutions.

CHAPTER 3

Sites of Learning in Eastern and Central Europe during the *Fin-De-Siècle*

The focus of this book will now shift and turn toward educational institutions and alternative sites of learning in Central and Eastern Europe during the *Fin-De-Siècle*. One can better contextualize Jewish intellectual life against the backdrop of schooling and alternative sites of learning such as the famous coffeehouse culture of Vienna. How Jewish intellectuals were schooled marks a trajectory of their thinking. If anything, by studying the oppressive nature of European schooling, one better understands what it is intellectuals were up against. Does a good scholar have to be oppressed in order to become an original thinker? I do not know the answer to this question, but it seems to me that oppression creates a certain kind of oppositional thinking. All of the Jewish intellectuals mentioned in this book are oppositional in their thinking, and they each contributed something unique to the canon. By elucidating a historical setting of educational institutions during the period in which these intellectuals lived, questions may be more adequately formulated about intellectual work and the ways in which intellectual work gets done in the midst of oppressive environments. These questions are not only important to ask in order to understand European Jewish intellectuals; similar kinds of questions can also be raised about American Jewish intellectuals who suffer similar kinds of oppressions.

Buber and Freud both attended Gymnasium. Anna Freud, however, did not. She attended what is called Realschule. Gymnasium prepared students for university work. Realschule prepared students for "real" world work. Steven Beller (1987) explains the difference between these European educational institutions in more detail.

> The Gymnasium was designed as the elite form of secondary education in Austria. It was not by any means the only form of secondary

> education; however, there were other forms, such as the Realschule, which provided a more vocational curriculum, the Burgerschule which offered a more rudimentary education, and the various specialist schools for the arts and crafts. (p. 47)

This Gymnasium's curriculum turned on the classics. An elite educational site, such as the Gymnasium, is by nature, a conservative one. Freud and Buber, who both attended Gymnasium, had a classical foundation upon which to build their ideas. In both cases, they drew on what they knew but branched off in totally different directions. What enabled them to branch out in their own ways? Does one need to have an orthodox background in order to become heterodox? Does one need to be steeped in a conservative curriculum in order to become an original thinker? Perhaps one needs to know the conversations of others before one can have one's own conversation or make that conversation one's own. The conservation of one's tradition is what a conservative or classical education entails. But one must not stay within the traditional conversation; one must break out of it somehow. Freud and Buber, for example, did contribute something new, in fact, they both changed the way people think about fundamental ideas around the notion of relation, for example. But how is it that people come to think original thought? This remains a question and a mystery. Can we really teach students to think? We certainly can teach them the classical canon or any canon for that matter. We can teach them to read texts carefully. But what about the act of thinking. Teaching how to think is a difficult problem. One is either a thinker or one is not. There is a point at which a teacher can no longer teach. Philosophical thinking probably cannot be taught; but thinking can be nurtured. Or thinking can push out from under oppressive atmospheres. It is hard to say how thinking occurs.

At any rate, it is important to note that the academic institutions in which Freud and Buber were schooled were by no means democratic places. In fact, Eastern and Central European high schools were known for being highly authoritarian and rigid, not unlike many American educational institutions. What is notorious about educational institutions, especially during the reign of the Habsburg Empire, is that they were oppressive. This becomes a curious fact against the backdrop of the genius of both Freud and Buber. How did they survive their schooling? How did they become original thinkers? What did tyrannical forms of schooling teach them about free thinking? What did a highly conservative curriculum do to foster the sense of original thought? It seems that a highly conservative tradition would not, in fact, foster original thinking. But that is not the case with either Freud or Buber. Eastern European schooling did not, in fact, encourage radical thought. And yet, many European intellectuals were schooled in traditional ways but broke out somehow to create new ideas and to write profound books. Is there a connection between tradition and radical thought? I leave these questions open ended. There are no clear answers to these paradoxes.

There is much commentary in the literature on the authoritarian aspect of European schooling and it is a well-known fact among historians. It is important to dwell for at least a moment on the authoritarian nature of these institutions, if only to ask whether American institutions are much different. Gary Cohen (1996) comments on the authoritarian nature of teaching in the Gymnasium.

> An authoritarian approach characterized the pedagogy as well. In Reichsrat debates on secondary education in 1891, for instance, Masaryk joined other deputies in criticizing the authoritarian character of the teaching, excessive emphasis on memorization, and the resulting neglect of independent thinking. (p. 216)

And yet, so many independent thinkers were schooled in this oppressive atmosphere. How did they free themselves from tyrannical schooling? How did Freud extricate himself? How did Bettelheim or Buber get themselves free from this oppression to create what they did? How did these thinkers manage to do what they did against this backdrop? Bettelheim (1989b) comments that "the Real Gymnasium ['s] . . . authoritarian nature offended me very much" (p. 104). In a short story titled "The Place I want to tell you About," by Joseph Roth (2002b), a character states that

> he hated the Gymnasium, he hated the rules, he hated the whole narrow-minded small town world he had been condemned to live in. So it was a huge relief to him, once he had scraped through his final exams. (p. 50)

Like Bettelheim and Roth, Stephan Zweig (1943/1964) comments on the wretched atmosphere of the Gymnasium.

> But because of their accurate arrangement and their dry formulating, our lessons [in Gymnasium] were frightfully barren and lifeless, a cold teaching apparatus which never adapted itself to the individual, but automatically registered on the grades, "good," "sufficient," and "insufficient." . . . It was exactly this lack of human affection, this empty

> impersonality and the barracks-like quality of our surroundings that unconsciously embittered us. (p. 30)

Harry Zohn (1943/1964) comments on Zweig's painful schooling experience as he states that Zwieg, "speaks for a whole generation—and more—when he criticized the authoritarian school system which produced stifling learning-mills and so conducive to psychological scars" (p. viii). In fact, Zweig (1943/1964) argues that the founders of psychoanalytic thought were products of this scaring atmosphere. Psychoanalysis might have been born of the trauma of schooling. Zweig (1943/1964) explains thus:

> We can look into the records of the psychoanalysts to see how many "inferiority complexes" this absurd method of teaching brought about. It is perhaps not chance that this complex was discovered by men who themselves went through our old Austrian schools. (p. 36)

European psychoanalysts were schooled didactically and yet the method of doing analysis with a patient is hardly didactic. Schooling wounds but need not kill. Still, authoritarian schooling must have impacted their work as analysts at some level. Did they learn to re-educate themselves? Deborah Britzman (2003) remarks, there is something about education that needs to be re-thought, there is an "after-education" about which Freud spoke. The "after-education" is, in a word, a new beginning. We need ironically to be re-educated after we are educated or *miseducated.* Maybe Jewish intellectuals had enough vitality and will to become re-educated. Otto Rank (1996) says, "as Freud himself always emphasized, it is rather an education, or reeducation [*Nacherziehung*]" (p. 113) that we need. Starting over is what psychoanalysis is all about. Perhaps we should start over with education generally as well. Re-educating oneself is no small task. One must build up a private library—both literally and in the psyche—that is uniquely one's own. One must learn from one's own archive. Once we leave school we must re-read and read again to re-educate. We must leave our teachers to teach ourselves how to read and what to read.

This is exactly what Anna Freud wrote about. Although Anna Freud did not attend the Gymnasium, she knew the horrors of the Realschule, and it is no wonder she wrote about school phobias in her works titled *Assessment of pathology in childhood* (1969a), *Child observation and prediction of development* (1969b), and *The child guidance clinic as a center of prophylaxis and enlightenment* (1969c). Children, Anna Freud argued, were not pathological but the schools in which they were placed were. Schools are madhouses just like universities. There is something sick about authoritarian schools; it is no wonder children act out. No wonder children develop school phobias. Schooling may be little more than punishment and humiliation. Peter Vergo (2001) comments on the "ossified" nature of schooling experience in the Habsburg Empire.

> One knew at the age of twenty exactly where one would be at the age of fifty. The unforeseen, the irrational were excluded; not only the administrative, but also the academic and cultural institutions ossified beyond any possibility of change. (p. 11)

Kafka's *metamorphosis*, Buber's notion of *surprise*, and Freud's turn to the *unconscious*, make more sense against the backdrop of an age of stagnation, predictability, and rationality. It is no coincidence that Anna Freud, Martin Buber, and Bruno Bettelheim, all products of Austrian schooling, turned their thoughts to pedagogy, curriculum, and education. Anna Freud knew from experience, exactly how *not* to treat children. Martin Buber is one of the few philosophers in the Western canon beside Rousseau and Dewey and A. N. Whitehead who had taken an interest in education. Buber might have turned to writing about education because of his horrible experience as a Jewish student in a parochial Christian environment. Buber fleshed out ideas about education in the context of the notion of the other. Bettelheim states that it was his horrific school experience that made him want to study education in connection with psychoanalysis. Bettelheim (1989a) remarks that "the first books that I found truly liberating were those which were critical of the existing educational system and this supported my conviction that there must be better ways to educate the young" (p. 104). Indeed, Bettelheim did find a better way to educate the young—through fairy tales! His horrific school experience led him into the realm of phantasy as a pedagogical tool. Bettelheim felt that fairy tales would allow children to connect to their unconscious worlds so that they might work through phobias and irrational fears. Bettelheim's (1989a) *Uses of enchantment* could be used as a school primer. Imagine that! Teaching young people through fairy tales is not a bad idea! Fairy tales open the imagination and allow children to think otherwise. Fairy tales teach much about the human condition. Fairy tales teach children to think analogically and metaphorically (for more on this see Mary Aswell Doll, [2000] *Like letters in running water: The mythopoetics of curriculum*).

Horrific schooling experiences forced many to find sites of learning elsewhere. Taking these alternative sites of learning—as a matter of study—demands attention. Alternative learning sites should not be thought of as a "mere" aside. Curriculum theorists argue that education happens not only on school grounds but on the edges of society as well (see, e.g., Munro, 1989). Many Viennese fled to the coffeehouses for intellectual stimulation, since schools were deadening places. Reading in coffeehouses has a sort of romantic appeal for young artists and writers. Freud, however, had little interest in the coffeehouse culture. Because he felt estranged in Austrian society, I suppose he felt that the Viennese coffeehouses were populated with enemies. Peter Gay (1998) points out that

> the Vienna that Freud gradually constructed for himself was not the Vienna of the court, the cafe, the salon, or the operetta. Those Viennas did very little to advance Freud's work. (p. 10)

Freud's site of learning was Bergasse 19, not the coffeehouse. Similarly, the University of Vienna was a troubling place for Freud because it was a hostile place for Jews as well as scholars who ventured into unknown academic waters. Bergasse 19, then, served as Freud's intellectual sanctuary. Certainly, the academy was not a sanctuary for him.

Like Freud, Hannah Arendt felt that she could do her best work outside of the academy. She had much disdain for academic philosophers who tended to be apolitical, as previously mentioned. Arendt had no patience for the university because, as she saw it, much time was wasted in trivial matters. And 90 percent of university life is trivial. Arendt had little patience for the cumbersome nature of academic trivia. She had little time to waste. She was an intellectual not a bureaucrat. Sometimes intellectuals do better work on the edges of experience, on the margins. Sometimes intellectuals are better-off working on their own, rather than in universities. Arendt worked better outside of the university.

Did the European university encourage reactionary politics during the *Fin-De-Siècle*? Certainly during Freud's time, the University of Vienna was reactionary. How could a man like Freud work in a place like that? How could any intellectual be bothered with climbing the ranks of the mediocre? Can an intellectual work inside of a hostile and reactionary institution and still produce important work? Are intellectuals really found inside of universities anymore? Perhaps intellectuals are not found inside the academy. Perhaps intellectuals are found at the edges of institutions or completely outside institutions. Buber speaks of the need to "recover" from institutions because they are inhuman, damaging places. How can we be good teachers inside of damaging places? Is pedagogy damaged always already? Or do we need to flee to the coffeehouse or our own houses on our own Bergasse 19? Do we need to flee the halls of the academy to rooms of our own? In a way, I would answer "yes." But in a way "no," unless one is independently wealthy. Who can afford to be an independent scholar anymore?

American and Austrian schooling are not so different from each other. Most teachers stand and deliver. This is what the administration encourages when undergraduate classes are filled to capacity; when 200 faces are staring at the professor what else is the professor to do? Students are expected to spit back material on scantron tests. Unfortunately, much of a university education is nothing more than rote memorization, especially at the undergraduate level. Educationalists have long argued that the American public educational system resembles prisons rather than schools. I argue that the university is a madhouse! Of course Foucault made similar implications throughout his work. But my thesis is more psychoanalytic, more experiential, more metaphorical than Foucault's. I argue that the dystopic university is the site of chaos and ungroundedness. At any rate, it is no wonder kids kill each other; it is no wonder that we live in the age of Columbine(s). Schooling perpetuates violence. Alan Block (1997) points out that

> it is a world in which the child's growth becomes a function of the violence that the world, in the form of its systems, exercises upon the child and in which the practice of education is centrally and actively complicitous. (p. 2)

With the advent of Columbine, educators need to re-think the entire American educational system. As Dennis Carlson (1998) argues, we need a radical decentralization of curriculum so as to empower teachers, to start all over again, to re-educate teachers to think for themselves so that they may teach children to think for themselves. That is what curriculum scholars attempt to do when teaching pre-service teachers. Imagine if we allow teachers to teach what they want. Imagine if teachers choose their own curriculum. Imagine if teachers explored ideas with children in ways that have little to do with standardization? What kind of citizenship would we be building then? Perhaps one that is better able to handle the challenges of living in a post–9-11 culture. Imagine if we were to educate children to really read, to read books that they loved, to find their literary companions, to learn to love reading, to learn to love studying. Imagine what kind of generation we would be building then. But that is not what American schooling does. Instead, American schools continue to perpetuate "the violence committed against children in the name of education" (Block, 1997, p. 3).

Part of the problem in Austria during the reign of the Habsburg Empire was that the curriculum was centralized. When the curriculum is state controlled, politics and education get tangled up in such a way that the demands of the State become the dogma of educational institutions. These two cultures, the culture of the State and the culture of the university, should be completely separated. But they are not. Public institutions bow down to the dictates of the State. The goals of a government should not be the goals of an academic institution. We saw the danger of a state-controlled curriculum in Germany during the 1930s. Hitler ordered German universities to clean house and fire Jewish professors. Martin Heidegger did just that as Rector of the University of Freiburg from 1933–1934. Heidegger also rid the university of Jewish students (Morris, 2001).

Austrian curriculum was, more often than not, dictated by reactionaries. Gary Cohen (1996) explains,

> During the first half of the 19th century, the Austrian government clearly expected its Gymnasium and universities to serve the interests of the state in training future government officials, clergy, and professionals. (p. 17)

Training government officials has little to do with the production of intellectual work. In fact, Austrian universities did not think it important to help scholars mature. Indeed, scholarly production was frowned upon. Alan Sked (2001) tells us that

> there was little regard for academic research. After all, Francis I had made the imperial position quite clear on that: "I don't need scholars but good citizens. . . . Who serves me must teach as I command. Anyone who cannot do that or who comes with new ideas can go or I will dismiss him." (p. 49)

Is this quote so astonishing I wonder? Curriculum gets dictated from the top down in public schools and universities are more and more under the critical eye of standardization fascists. Should it come as a surprise that institutions of higher learning are madhouses? It is not uncommon in universities to ask scholars to do the bidding of the administration, while ignoring the importance of scholarly work. This is mad. What does higher learning really mean? Is higher learning a code word for shutting down intellectual engagement? In all institutions, I suppose, there is a shutting down of thought because one must serve the institution. One must work for the will of the master. There is something about institutions which is inherently sadistic. One way to control citizens is to squash them intellectually. Scholars were much feared by the dictatorial regime in Austria. The power of books obviously threatened the emperors. Hence, in Austrian institutions, academic freedom was sorely lacking. Gary Cohen (1996) explains the troubles professors had during the early part of the nineteenth century.

> Leopold Von Hasner (1818–91)—law professor, state official and German liberal minister of religion and instruction from 1867 to 1870—was born in Prague, the son of a government official. He entered the Old Town Gymnasium in 1826 and studied in the Prague law faculty between 1836 and 1842. Looking back, Hasner recalled how strong the efforts to regulate academic studies, even to the point of stifling scholarly inquiry, in order to stop the spread of dangerous liberal ideas. The Jewish poet Ludwig August Frankl (1910–94) came to Vienna medical faculty in 1828 after finishing studies in the lyceum of Litonysl, Bohemia. Nearly all his professors in Vienna were engaged in research and writing, but he complained that their teaching consisted of imparting information and drilling the students without stimulating any independent inquiry. (pp. 18–19)

Intellectual experimentation has always been a threat to imperial and non-democratic institutions and cultures. One way to control liberal thinking is to censor it outright and outlaw it. This is exactly what Austrian officials did. Historians comment that during the reign of Habsburgs, a spy system was set up to censor all kinds of publications from newspapers to scholarly monographs. But censorship was not watertight. Gary Cohen (1996) explains,

> Austrian censorship stopped the printing of liberal and radical ideas inside the monarchy, but the empire's subjects could still read them in newspapers and books brought in from some of the German states. In fact, the Austrian government had less control over society than is often assumed. (p. 20)

William Johnston (1972) concurs. He speaks of the "sloppiness" of press censorship. Johnston explains,

> Nothing illustrated so well the sloppiness of bureaucracy as the manner in which it handled censorship of the press. Each morning preliminary copies of every paper were rushed to the censor, who

> might order any story confiscated. In its place an empty space bearing word Konfisziert. (p. 49)

Austrian secret police spied on everyone. Alan Sked (2001) tells us,

> Hence Robert Justin Goldstein . . . has written: "The Austrian secret police were the most notorious during the 1815–1860 period. . . . Huge numbers of informants—especially in such occupations as servants, prostitutes, waiters and doormen—were hired to report to the police on the activities and conversations of Austrians." (pp. 45–46)

It is astonishing that spying became part and parcel of Austrian culture long before the rise of the Nazi regime. Spying was nothing new to Austrian citizens. Spying probably seemed natural to them by the time Hitler appeared on the scene. Not surprisingly, Sked (2001) points out that Jewish writers were the most censored of all. Jews were the most feared throughout the empire. Today one must wonder about the Patriot Act and all of its implications. There are certainly uncanny parallels between Austrian culture and the current culture of America. This is extremely alarming.

Jews were associated with liberalism and conspiracy. Yet, Franz Joseph protected the Jews. In fact, some argue that the golden age of Jewry existed under the reign of Franz Joseph. Still, Jews were under suspicion and were the focus of spy intrigues. Against this backdrop, one understands better why Freud had such a hard time being accepted by his Austrian compatriots. Austria was not a culture, by and large, open to new ideas. Austrian universities were not places of experimentation. Large segments of Austrian culture were conservative. Of course much of the art culture, as Peter Vergo (2001) has pointed out, was experimental. But not all segments of Austrian culture cared for Klimt, for example. In fact, some Austrians saw Klimt and Schiele (probably the two most well known artists of the *Fin-De-Siècle* in Austria) as criminals and perverts. Likewise, Freud's work was received with great horror.

Despite censorship, Austria was—at least during Freud's lifetime—a place of tremendous intellectual creativity. Scholars are conflicted as to why or how this was possible. Some suggest this extraordinary creativity was coincidence, while others suggest it was not. I argue that Jews, in particular, created remarkable works because they were oppressed as Jews and needed an intellectual outlet to express the tensions they felt living in such a hostile climate. Of course, some of the great minds of Austrian culture were not Jewish. But by and large, many of them were. There is something about being a stranger in a hostile culture that makes people write differently, write oppositionally. Jews were constantly insulted and belittled by anti-Semites. It is important to understand that unlike American culture, Austrian culture has always been a hotbed of *overt* anti-Semitism. Americans have little understanding of the harshness of anti-Semitism experienced by Austrian Jews. Some Americans may be anti-Semitic, but I do think most hold their tongue more than Austrians.

Jewish intellectuals in Austria were perceived as "too Jewish" because they were scholars, "bookish" and liberal. Historian Robert Kann (1980) tells us that a Christian Socialist Austrian politician named H. Bielohlavek declared,

> "I am fed up with books, you find in books only what one Jew copies from another" or Lueger's own publicly expressed opinion that the liberal scholars "should shut up, until one of them could invent artificial grass which a real cow could eat." (p. 435)

What is astonishing is that public figures spoke *publicly* like this! Lueger was mayor of Vienna, the first mayor in fact, who ran on an anti-Semitic platform (Pauley, 1992). Recall, Anna Freud's shocking statement about the way people talked to and about Jews in Austria. Because of the Jewish love for learning, many Jews filled the lecture halls at the University of Vienna and many Jewish students were enrolled at the Gymnasium, not the Realschule. Of course the reactionary culture did not make it easy for Jews at university. Freud had a hard time too, as I have pointed out. Peter Gay (1998) suggests that Freud's Jewish identity began to become important to him while at the university because of the anti-Semitism that he encountered there. Gay remarks, "It was not without reason that Freud should date his particular Jewish self-awareness to his years at the University of Vienna, where he began his studies in the fall of 1873" (p. 15). One becomes what one is because of social pressures. The hostility Freud encountered forced him to grapple with his Jewishness. When one is oppressed, one is forced to grapple with one's ethnicity. White people think they do not have a color, or think of themselves as white, because they are not forced to do so since they are part of mainstream culture. But because Jews are pushed into a corner, as it were, they are continually forced to think about what it means to be the other in the context of hostility. The stronger anti-Semitism becomes, the more some Jews become wedded to their Jewishness. Another alternative is to convert or just ignore discrimination and prejudice. But many people feel more Jewish when anti-Semites push them into a corner.

It is astounding that Jews attended university at all against the backdrop of a reactionary and violent student body at the University of Vienna. Peter Gay (1998) explains,

> Demonstrating their proverbial appetite for learning, they [Jewish students] poured into Vienna's educational institutions and, concentrated as they were in a few districts, clustered in a few schools until their classes resembled extended family clans. During the 8 years that Freud attended his Gymnasium, between 1865 and 1873, the number of Jewish students there increased from 68 to 300, rising from 44 to 73 percent of the total school population. (p. 20)

The mere fact that Jewish students enrolled in university enraged anti-Semites. How Jewish students dealt with anti-Semitic harassment will never be known. Harassment, though, did not merely consist of words. What began as name-calling ended in death camps. What started as exile ended in annihilation.

Those who were lucky enough to escape Austria found homes mostly in America and Britain. During the rise of the Third Reich, many Austrian and German intellectuals emigrated to the New School for Social Research. One of the great projects that came out of the New School was the study done on the authoritarian personality headed up by renowned psychoanalysts and critical theorists, Adorno, Horkheimer, Fromm. No wonder. How could Hitler *not* have an authoritarian personality growing up in Austria? Perhaps this thinking is too simplistic. Not all Austrians were authoritarian, but against the backdrop of the long history of authoritarian institutions, it seems understandable that scholars would study this topic. Hitler's regime seemed a natural outgrowth of Austrian culture. Hitler was not an eccentric by any means. He was, if anything, an ordinary Austrian. (See, e.g., Raul Hilberg 1961/1985; Christopher Browning, 1998; Daniel Goldhagen, 1997 on the notion of *Ordinary Germans.*) Hitler was mediocre, loud-mouthed, violent, and reactionary. There was nothing extraordinary about him.

The exile of Jewish scholars from the Austrian educational system occurred long before Hitler arrived on the scene. European Jewish intellectual culture was completely and totally destroyed by Hitler. By the time Freud left Austria in the late 1930s, nearly all the Jewish psychoanalysts fled Austria, as I mentioned earlier. Some were not so lucky. Bruno Bettelheim spent time in a concentration camp.

Freud's sisters were murdered in death camps. Vienna was home to many Jewish Nobel prize winners, but they were all exiled. Imagine that! Universities did not want anything to do with Jewish Nobel prize winners because they were Jewish. Egon Schwarz (1999) tells us,

> All Nobel prize winners who were active in universities were dismissed and hundreds went into exile. None of the 14 who made up the Vienna Circle of philosophy ever returned. Three thousand physicians fled Austria. The renowned Vienna School of Art History was exiled and largely incorporated into the English Warburg Institute. The architectural avant-garde fared no better. (p. 101)

One has to wonder what happens when a culture is so decimated. Who are the great Austrian intellectuals of today? Is there another Sigmund Freud? I think not. Is there another Anna Freud? I do not think so. These people cannot—and will not—ever be replaced. Damage that has been done cannot be undone. Austrian culture will always be stained with the memory of the dead. The streets of Austria are haunted and will be haunted until the Austrians decide to finally look at what was done. Historians feel that Austrians have not done this, not, at any rate, as the Germans have. The Austrians are still not willing to grapple with their past.

The reactionaries who filled the universities during the 1930s managed to completely destroy the *Fin-De-Siècle* culture that made Vienna so famous. Bernard Handlbauer (1999) argues that the

> expulsion of Austria's intellectual elite in the 1930s was the most drastic rupture in Austria's cultural history of this century. By the end of 1938, the heritage of the Fin De Siècle, with its innovations in architecture, art, literature, medicine, music, philosophy, and psychology, had virtually disappeared from Austrian soil. (p. 109)

Austrian culture can never recover from this gaping hole, this horrific void. Austria is ominous even today. One visits Freud's house on Bergasse 19, but there is nothing left in the house except photos. At a music museum a small Holocaust display is installed as an afterthought, an aside, a mere blip on the screen of history. The Holocaust monuments around the inner city are so small that it would be easy not to notice them at all. The portrait of Gustav Mahler hangs in the famous Viennese Opera House but seems surrounded by ghosts. This portrait is a lie. Many Austrians still claim that they were "invaded" by Hitler's army and were not responsible for what happened to the

Jews. The lie of history. The truth is they welcomed Hitler with open arms and killed more Jews than did the Germans.

Many people fantasize that Vienna is merely coffeehouses and operas, lippizzaners, sachertortes, and yodeling. Austria is the *Sound of Music*, Julie Andrews, and kitsch. Yet there is more to it than that of course. Whose Vienna is the question. Book titles suggest that there is Freud's Vienna (Bettelheim, 1989b), Wittgenstein's Vienna (Janik and Toulmin, 1996) and Hitler's Vienna (Hamann,1999). Vienna is a confusing study, for it is many things to many people. One might be both attracted and repulsed by Viennese culture. Vienna is the home of Freud and the Holocaust. Vienna is the place where great intellects were born and horrific mass murders were systematically carried out. Vienna is the site where Hitler was born. Bruce Pauley (1992) reminds us that "half of the crimes associated with the Holocaust were committed by Austrians even though they comprised only 8.5 percent of the population of Hitler's Greater German Reich" (p. xix). Not only this, Austrians, in many cases, were more vicious than the Germans in their brutality. Gunter Bischof (1997a) argues that

> it is well established by now that the "Vienna model" of expropriation of Jewish property through "Aryanization" and the forced emigration of the Jewish community after the Anschluss encouraged and radicalized anti-Jewish policies in the Altreich during the course of 1938 culminating in the infamous November pogrom ("Reichskristallnacht"). (p. 271)

Walking through the streets of Vienna today is a sad experience. It is overwhelmingly beautiful and uncannily creepy. It is the home of Freud and the site of bloody brutality. Vienna is a city of torment, a city that has forgotten her past, a city of trauma and nightmare.

Anti-Semitism and Otherness

What was it that compelled Jews to feel like strangers? Why is the pattern of strangeness so clear? People experience Otherness because they are forced to live outside the mainstream culture, they are forced to live on the margins of society. Living on the margins creates an experience of Otherness, of strangeness. Otherness for Jews is caused by experiencing anti-Semitism. Austrian Jews had to contend with anti-Semitism. Austria, in particular, was perhaps the most anti-Semitic culture in Europe, especially after the fall of the Habsburg Empire, although historian Bruce Pauley (1992) contends that anti-Semitism in Austria can be dated back to the tenth century. Jewish intellectuals had to cope with this extreme form of hatred. Writing and creating works of art are ways of coping with being othered. If Jews attempt to assimilate to Christian culture, painful emotions erupt. If Jews do not assimilate, this too becomes a painful experience. Being "out" as a Jew in Austria must have been a horrible experience during the war years. And yet it is this pain that might drive creative work. Péter Hanák (1998) comments that

> the well-known social psychologist, Kurt Lewin, acknowledges that assimilation to gentile society plunges Jewish intellectuals into marginality, generating psychic tensions and conflicts. These are socially and psychologically harmful, he argues, but may be beneficial for creative work. (p. 176)

Psychologically, Jewish thinkers have to negotiate a sense of Otherness within because of external pressures. One must question one's own worth, one's own sense of well being amidst marginalization and hatred. This questioning creates a psychic state of discomfort and out of this discomfort comes oppositional thinking. Could Jewish intellectuals become oppositional thinkers if they do not work out of states of melancholy, exhaustion, or despair? Or are these questions irrelevant? Does it matter what makes intellectual life work?

These questions are important to ask, even if there are no answers. Of course one can do good intellectual work without suffering, without despair and anxiety. Certain kinds of work, though, spring from being othered. As I previously mentioned, oppositional thinking does not come out of nowhere. However, I do not want to make the case that one must suffer to do good work. But I do want to suggest that suffering makes for a certain kind of work. Certainly, Freud suffered and felt that this was necessary for his production. But not all creative intellectuals suffered as he did.

Jewish intellectuals have historically been discriminated against and this is cause for suffering. As I mentioned earlier, Freud had trouble getting promoted in the academy because he was a Jew. Buber's library was destroyed by the Nazis because he was considered an "arch-Jew" (Friedman, 1991, p. 215). Maurice Friedman (1991) tells us,

> During the ravages and brutality of the Kristallnacht in November 1938, when Buber had already emigrated, the Nazis destroyed all the furniture in the house in Heppenheim, which still belonged to Buber, and three thousand volumes that still remained in his library, and then

> sent him a bill demanding that he pay them twenty-seven thousand marks! (p. 210)

Buber had to pay for the Nazis's destruction of his own library! Jews also had to pay for the damages done during Kristallnacht when Jewish businesses and homes were ransacked and totally destroyed. Windows in Jewish shops shattered. Furniture in Jewish homes ripped apart, dishes broken, paintings slashed. A violent free for all by the Nazis. The world looked on in horror, but did nothing about it. Back to Buber for a moment. Buber's writings certainly outlasted what damage the Nazi's thought they did to their "arch-Jew" (1991, Freidman, p. 215). The Nazis thought they could destroy Buber by destroying what he loved most. But they were wrong. Buber was not destroyed. Buber is still with us today through his writings.

Buber was a religious Jew and wrote about his religiosity. For this, the Nazis hated him. Religious Jews had historically been represented as aliens, as Others, as outsiders, parasites, vermin, rats, and even vampires! Brigitte Hamann (1999) contends that George Schonerer in the late 1800s (the anti-Semitic Austrian politician upon whom Hitler modeled himself) declared, "Like vampires, . . . [the Jews wanted] to suck their vital force from the strength of the Aryan peoples." And: "Every German has the duty to help eliminate the Jews as much as he can" (p. 246). What is astonishing is that some Austrians were calling for the annihilation of Jews some forty to fifty years before the Holocaust. I often wonder what could possibly bring about so much hatred? The ridiculous nature of Schonerer's bombast might be dismissed. But in retrospect, these words become eerily prescient in light of the Holocaust. The spoken word, the written word, and the act are not totally separate. This is the point Péter Hanák (1998) makes, while uncovering disturbing facts in European newspapers. Hanák claims,

> [T]he Jew represented something anomalous or demonic in which the spirit of witchcraft and destruction dwelt. This was converged in the distorted ears, nose, and lips and the obligingly servile, cynically shallow, or hair-raisingly rapacious grin that all anti-Semitic caricatures wore. . . . That figure, with his thick wallet and purse stuffed with gold, marched on from those German, Austrian, and Hungarian papers to the pages of the Nazi Sturmer and other fascist papers—and from there straight to the death camps of the Second World War. (p. 49)

Fantasies about Jews tell us more about anti-Semites than about real Jews. The Jew gets constructed through the projections of the anti-Semite. People split off fears and aggression and project them onto scapegoats. Not being able to psychologically manage one's inner demons is the root of all prejudice. These inner demons, if not dealt with, worked through and integrated in the self, get projected outward onto a scapegoat.

Anti-Semitism has certainly been around for a long time. Yet, what is shocking is the continuity of anti-Semitism over time, even though the forms of anti-Semitism change. Even though we know that unresolved psychological conflicts get projected out onto scapegoats, scholars still do not fully understand why hatred persists—even when conflicts are somewhat resolved. What is it about rage? What is it about murderous rage? There is always already hatred and sadism in every culture. In fact, Freud said that hatred is older than love. But why? Perhaps Melanie Klein is helpful here. Perhaps these sadistic impulses spring from early childhood. Hate is an archaic structure in the psyche. Hate is old in that it begins when one is a child, for whatever reason. Hate is hard to undo because it is so psychologically old. One does not just wake up one day and hate. Hate is in the heart for years. And hate needs an object.

Saul Friedlander (1999), in a brilliant piece titled "Europe's Inner Demons: The 'Other' as Threat in Early Twentieth Century European Culture," tells us,

> In Goebbels' diary entry, the otherness of the Jew is absolute as the difference between the beholder's world and the "synagogue," between "human beings" and "animals." The threat represented by the Jews as the quintessential Other is illustrated by increasingly extreme metaphors. First, the "brutality" of the Jew's beast-like life provokes such horror that "one's blood freezes"; then, the merely horrifying brutishness turns into mortal danger: the Jew as pestilence and disease. (p. 211)

Nazis hated both ethnic and religious Jews. Ironically, assimilated Jews terrified Nazis more, because they could not tell who was Jewish and who was not. The more Jews tried to blend into European culture, the more hated they became. One cannot contain what one cannot see. One cannot contain the "Jewish conspiracy" if one does not know who is Jewish. Saul Friedlander (1999) suggests that whereas in general the Other's most threatening aspect

> seems to reside in an identifiable difference, the most ominous aspect of the Jewish threat appeared as related to sameness. The Jews' adaptability seemed to efface all boundaries and to subvert the possibilities of natural confrontation. (p. 213)

That assimilation creates more problems for Jews is hard to understand. Jews become invisible and blend in with the crowd and are harder to see. Jews could be next door neighbors! Perhaps it is the close proximity to the Other that anti-Semites fear. As long as Jews are at a distance they can be separated out from the masses. But assimilated Jews are hard to define and even harder to control. When it seems that Jews blend into the crowd their movement cannot be easily traced. Peter Gay (2002) comments that

> a Jew, anti-Semites argued, could never find a true home. If he persisted in his traditional piety, perhaps even in his traditional garb, he was only making a highly visible statement about his incurable otherness; if he tried to assimilate by changing his name or being baptized, he was only making a transparent attempt at camouflage, unwittingly revealing his characteristic Jewish cleverness. Once a Jew always a Jew, the racists proclaimed, always alien, always dangerous. (p. 115)

Jews are a people of the Diaspora. Diaspora means being scattered against one's will. Jews could not "find a home" because they were kicked out of nearly every European country over centuries. Whether assimilated or not Jews lose. Jews are always already "pathological" and evil—according to anti-Semitic lore. Jews are always, at bottom, the root cause of world destruction in the eyes of anti-Semites. We cannot be hated enough. Hatefulness is curious. Of course I confront them head on, but it seems to do little.

Trying to understand what anti-Semitism is, is no easy task because of complications associated with assimilation and difference. I have treated anti-Semitism in some depth elsewhere (Morris, 2001). Here, I only want to offer a few more comments on anti-Semitism to better contextualize my study of European Jewry and education. Marin (1987) argues that anti-Semitism must be treated historically. He states, "The basic starting point . . . is a rejection of ahistorical-mythical interpretations of anti-Semitism as a timeless 'eternal' phenomenon" (p. 216). The notion of anti-Semitism changes with the tide of history; its form and content change over time. Anti-Semitism is also place-bound. That is, it tends to emerge in particular places rather than others. For example, anti-Semitism does not have a history in Italy, even though some Italians collaborated with the Nazis. On the other hand, Austria has had a long history of anti-Semitism. America does not have a history of anti-Semitism—like Europe—but many anti-Semites live in the United States, and certainly, there are broad segments of the population who are anti-Semitic. The face of anti-Semitism has now shifted with the Iraqi war. Muslim anti-Semitism differs from Christian anti-Semitism, because of its history and politics. Anti-Semitism is found both on the left and on the right of the political spectrum.

Anti-Judaism, some argue, is the precursor to anti-Semitism. Anti-Judaism appears on the scene with the advent of Christianity. Some argue that anti-Judaism dates back even before Christianity. Scholars tend to separate the terms anti-Judaism and anti-Semitism because anti-Semitism, as a term, did not appear until 1879 when it was coined by the anti-Semite Wilhelm Marr (Poliakov, 1974). With the rise of phrenology, racial anti-Semitism becomes a "science." Phrenology is actually a pseudo-science. The purpose of this so-called science was to separate the Aryans from the Semites. Phrenology was used to oppress.

There are many differing kinds of anti-Semitism. Cultural anti-Semitism and economic anti-Semitism, for example, are forms of hatred that differ from religious anti-Semitism or what some call anti-Judaism; political anti-Semitism is perhaps the most dangerous kind. Political anti-Semitism was born in Austria. Austria was the first country to embrace official anti-Semitic policy (Pauley, 1992). Bruce Pauley (1992) suggests that there is also "religiously inspired" anti-Semitism that crosses over into political anti-Semitism. But the most shocking anti-Semitism is student anti-Semitism. Bruce Pauley (1992) tells us that

> like Germany no other group in Austria was so racially, passionately, and violently anti-Semitic as students of university age. Jewish students were frequently attacked and anti-Semitism was so common that it was almost taken for granted. This anti-Semitism was tolerated both by sympathetic administrators and, until 1930, by the tradition of academic freedom. (p. 98)

Like Pauley's study, Peter Pulzer's analysis of student anti-Semitism in Austria troubles. Pulzer (1988) says, "Student anti-Semitism, stronger in Austria than in Germany, was predominantly nationalist" (p. 221). In fact, Pulzer claims that "Nationalism had become the main driving force behind anti-Semitism" (p. 237) in the early part of the twentieth century. Students' prejudices are already hardened by the time they reach college age.

Although anti-Semitism has a long and enduring history, its appearance and manifestation continually change depending upon its sociopolitical context. Marin (1987) states that contemporary anti-Semitism has a different face than it did in the past.

> In fact, today's enduring "post-Fascist" anti-Semitism seems to differ as much from its predecessor as the "modern" anti-Semitism that started in the 1870s differed from the ancient, religious "anti-Judaism" that dates back to the Middle Ages. (p. 217)

Whatever anti-Semitism is, it is continually shifting and changing because it is a concept that is historically and socially constructed and reflects the nation and place in which it is contextualized. What is striking, though, is its virulence. One need only look to popular culture in America to see that anti-Semitism is alive and well.

The Fall of The Habsburg Empire

Austria is no *Sound of Music.* If anything, Austria has had a bloody history with which it has yet to comes to terms. Historians often argue that the Third Reich was not inevitable, however. It was not inevitable that after the *Fall of the house of Habsburg*, as Crankshaw (1963) puts it, the Holocaust was fated. Yet, something sociologically and culturally did happen between the Great War and 1930, the year Hitler was elected as Chancellor of Germany. Something happened. Some unthinkable floodgate of hate opened to allow 66,000 Jews in Austria to disappear, to be brutally annihilated. Six million total annihilated. How could it be that such a cultured society could turn so monstrous? Austria was not always so monstrous though. As I pointed out earlier, under Franz Joesph's reign, minorities were protected from brutality. Franz Joseph somehow managed to hold together competing ethnicities, or so he thought. George Strong (1998) comments,

> A study of the political culture that took hold of Austria-Hungary after 1850 has relevance for Americans today in that the history of the Dual Monarchy may be viewed as a failed experiment by a state that attempted to sustain itself by reconstructing itself as the basis of its newly discovered cultural diversity. (p. 3)

The idea of cultural diversity is nothing new in academic discussions. Habsburg was vastly diverse. And there was certainly a lot of discussion about this issue. Competing ethnicities lived side by side in Central and Eastern Europe, not always harmoniously. Thanks to Franz Joseph, these diverse peoples were able to live in close proximity without killing each other. But these competing ethnicities simply could not get along after Franz Joseph's empire crumbled. Historians argue that the

Habsburg Empire crumbled for many reasons. One was Franz Joesph's incompetence. Alan Sked (2001) claims that Habsburg crumbled because Franz Joseph insisted on maintaining a rigid "status quo" (p. 274). Allan Janik and Stephen Toulmin (1996) argue that a "petrified formality" (p. 37) kept in check complete and utter chaos. Competing ethnicities eventually tore Central and Eastern Europe apart. Some argue that the empire was too oppressive; the ground swell of the oppressed became too great for the emperor to handle. A. J. P. Taylor (1948/1976) argues that the collapse of the empire had directly to do with Franz Joseph's' lack of responsibility. Taylor remarks, "Lacking faith in his peoples, he felt no responsibility toward them and made concessions from fear, not from conviction. As a result, he became the principle artificer of the collapse of the Habsburg Empire" (p. 78).

The dual monarchy was highly oppressive, even though Jews gained civil rights under Franz Joseph. But still, the empire was nonetheless imperial and oppressive. Time and time again it is commented upon that the empire's seeming orderliness masked an underlying chaos. Nothing was ever out of place and everyone knew their place. But people did not like this. People like freedom and the empire was not a free society. The culture of Austrian schooling certainly mirrored the larger authoritarian culture.

In a grand move of reversal, after the *Fall of the house of Habsburg* (Crankshaw, 1963), the oppressed minorities, with the exception of the Jews, became the oppressing majority! Crankshaw (1963) puts it this way, "It was soon found that the new master-nations—Czechs, Poles, Serbs, Italians—had learnt all the techniques of oppression. Then came Hitler" (p. 3). That the oppressed become the oppressor is not an unusual phenomena. Victims of child abuse sometimes become abusive toward their children. The colonized sometimes become the colonizers. One thing that non-Jewish minorities of the Habsburg Empire had in common was that they all hated the Jews. In Europe, there was no group of people more hated than the Jews. Péter Hanák (1998) reports,

> One can also read of a hierarchy among the minorities. The Hungarian nobleman had little liking for the urban German burgher, but both alike despised the Slovak, Rumanian, and Serbian tradesman, artisans, and peasants, all of whom joined them in a common hatred toward the Jews. (p. 53)

Sander Gilman commented at an MLA meeting in New Orleans in 2001 that the Jews are the most Othered group of minorities worldwide.

There is plenty of hate to go around among other groups too. Today, competing ethnicities across the globe cannot seem to manage difference. The hatred that differing ethnicities feel toward each other has not abated. I do not mean to essentialize here, but generally speaking many Germans hate the Turks (the new German scapegoat), many Muslims hate Christians, many Serbs hate the Croats, many Palestinians hate Israelis, many Catholics hate Protestants. City mouse hates country mouse. Northerners hate Southerners. Everybody hates the Jews. I wish to qualify these hatreds by deliberately using the word "many" as cumbersome as it is because it is not true that everybody hates. There are good people in the world too. The problem is that many people harbor hatred in their hearts. The hate just never seems to end. Sometimes I wonder what good courses in multicultural education do? Students are hardened before they enter the classroom and seem only to grow more firm in their convictions after leaving the course. Xenophobia, racism, religious intolerance, and nationalism become problematic, especially when combined. And it is the case more often than not, all of these "isms" do get connected up. In other words, someone who is xenophobic, is probably racist, anti-Semitic, and homophobic.

Historians claim that one of the major problems in Austria has been issues related to nationalism after the fall of the Habsburg Empire. After the Habsburg Empire crumbled, nation-states rose, and it was the rise of the nation-states that created problems for Jews. It is ironic that in the eyes of anti-Semites Jews could not become citizens of nations but were expected to. Marsha Rosenblit (2001) remarks that with the rise of the nation-states people then,

> demanded that the Jews identify with the dominant state. . . . Unfortunately, though, anti-Semitism flourished in most of these states, and radical anti-Semites denied that Jews could ever become part of the nation in any sense at all. (p. 10)

The Jews became everybody's scapegoat. Outbreaks of influenza were blamed on the Jews. The Great War was blamed on the Jews. Communism was blamed on the Jews. Jews were blamed for plagues. The stock market crashes (of both 1873 as well as 1929) were blamed on the Jews. Peter Gay comments (2002),

> In May 1873, there had been a spectacular crash on the stock exchange, with baneful consequences for banks and for investors all across Europe, Schnitzler's father among others. This gave anti-Semites the right, or so it seemed to them, to blame the Jewish speculators for the

> vagaries of the Austrian money market, to call for the ouster of Jews from university fraternities and public employment. (p. 116)

What is disturbing here is that these calls for the ouster of Jews from public life did not begin with Hitler's reign of terror in the 1930s. Austrians, in particular, had been calling for the elimination of Jews from public life for decades before the Holocaust. It was just a matter of time before Austrian public officials actually carried out these oustings.

Ironically, it was not the era of empires that brought about the worst forms of anti-Semitism but rather the *democratization* of Eastern and Central Europe. Democracy brought about the gas chambers and death camps. Hitler was democratically *elected* as Chancellor. Astonishingly, Peter Pulzer (1988) argues that "anti-Semitism is unthinkable without democracy" (p. 287). The democratization of Eastern and Central Europe meant that the anti-Semite was now freed to hate the Jews. The democratization of Eastern and Central Europe meant the freedom also to act on this hatred. Bruce Pauley (1992) explains,

> Freedom of speech and freedom of assembly also meant freedom to shout anti-Semitic slogans and to hold anti-Semitic demonstrations. Anti-Semitism, in fact, flourished in all the new democracies of East Central Europe after 1918, with the exception of Czechoslovakia and Yugoslavia. (p. 73)

One does not usually connect democracy with brutality; one usually associates dictatorship with brutality. But it was *because* people had the right to vote, the right to voice their opinions, the right to make a choice for Hitler that the eventual annihilation of the Jews came about. Against the backdrop of history, one must wonder about the potential problems of newly formed democracies. Iraq gives Jews pause.

After 1888 Austria gave voice to major political parties (the Christian Socialists, the pan-Germans and the Social Democrats) all of which, to one degree or another, held anti-Semitic views (Pauley, 1992). The Social Democrats, a party to which many Jews were loyal, were the least anti-Semitic. It alarms that anti-Semitism was present even in the most liberal of all Austrian political camps. The most anti-Semitic of the three parties was clearly Christian Socialism. George V. Strong (1998) contends,

> In its predominantly popular, lower form, Christian Socialism stands condemned in history for its crude pro-Habsburg, pan-German outlook, heavily laced with anti-Semitism because dislike of Austrian Jewry

> was linked in large measure to antimodernism—that is, to popular hostility toward capital and large industry. (p. 30)

Interestingly enough, these were the reasons Heidegger did not like the Jews (Morris, 2001). These are also the reasons behind Russian anti-Semitism. Many Russians hate Americans because they associate America with capitalism and capitalism with cosmopolitanism and cosmopolitanism and capitalism with Jews. Of course Russian anti-Semitism differs from Austrian anti-Semitism because of its own unique history and sociopolitical context, but still one can trace some common themes here.

In Austria, long before Hitler, the infamous Mayor Karl Lueger became the leading voice for the Christian Socialists. Lueger became mayor in 1888. Hitler became chancellor of Germany in 1930. Lueger was Hitler's mentor and model. Brigitte Hamann's (1999) book title is apt: *Hitler's Vienna: A dictator's apprenticeship.* Hitler learned everything he knew from the Viennese. Brigitte Hamann (1999) remarks,

> The mayor was not squeamish: when a liberal Jewish deputy protested in the Reichsrat against the incitement of people to anti-Semitism, Lueger replied that anti-Semitism would "perish but not until the last Jew has perished." When an opponent recalled Lueger saying, during a mass rally, that he did "not care whether the Jews are hanged or shot," Lueger unfazed, interrupted him to correct him: "Beheaded. I said." (pp. 286–287)

What is astonishing is that today in Vienna a street is named after Lueger. Are the Austrians proud of Lueger's legacy or are they merely forgetful? That the street is named Lueger is symptomatic of Austria's refusal to own up to deeds done during the 1930s and 1940s.

In sum, Jewish intellectuals, against the backdrop of anti-Semitism, might develop oppositional ways of thinking. Austrian schooling and Austrian institutions of higher education were hotbeds of anti-Semitism. What Jews actually learned in school was how to survive a continual onslaught of hatred. What Jews learned in school was that they had to create their own cultures of learning, their own Bergasse 19s in order to do intellectual work. School was not the place where great minds flourished. But school created the backdrop against which intellectuals began thinking about thinking otherwise.

PART II

Madness

CHAPTER 4

Madness as a Trope of Otherness

The first part of this book examined the trope of Jewish intellectuals as a site of Otherness. Because of the persistence of anti-Semitism, Jewish intellectuals have always had to do their work against the backdrop of hatred. Jewish intellectuals have had a pattern of engaging in oppositional thinking as a direct response to the ways in which they historically have been othered.

The second part of this book begins by examining a more psychological aspect of Otherness. The *experience* of Otherness is examined by fleshing out the concept of madness. Exploring the notion of madness may enable scholars to gain a better understanding of the way extreme states of Otherness *feel*. Madness will be examined psychoanalytically and phenomenologically.

I want to explore—in a deeply psychological way—experiential states of Otherness and later connect these states in a metaphorical way to what I call the dystopic university, which I will treat in the last part of the book. More specifically, the larger goal of the book is to eventually raise questions of how Jewish intellectuals survive chaotic feeling states that are experienced, while working in dystopic universities.

Subjectivity and Altered States

I begin this chapter by exploring subjectivity and altered states. If scholars want to understand what Otherness feels like, phenomenologically and psychoanalytically, they might begin by examining their own dreams, phantasies, wishes, and desires. This is no easy task because all of these emotions are slippery, and they are all what I would consider boundary states.

Karl Jaspers (1932/1971) writes about "boundary situations" (p. 192) and of "foundering" (p. 192). Radical forms of subjectivity founder between the conscious and the unconscious. Foundering means becoming Other to oneself. Becoming Other to oneself means becoming more attuned to one's lifeforce, one's lifework. Becoming Other profoundly shifts subjectivity toward a Being-on-the-edge. Being-on-the-edge requires the risk of what Jaspers calls "foundering." Jaspers (1932/1971) contends that

> as a phenomenon, foundering remains contradictory. The solution is not known. It lies in being, which remains hidden. The man who has really climbed the existential steps in his own proper fate comes up against this being. It cannot be presupposed. There is no authority that could be its administrator and mediator. Its eye is upon him who dares approach it. (p. 196)

Foundering is that phenomenological state of Being that one cannot grasp. The earth moves to and fro; one's life is neither here nor there and one's path is that of no path. Foundering at boundary situations, then, would be for Jaspers the place where thought begins. Boundary situations, philosophically, are not clear.

Foundering is much like the notion of ambivalence. Ambivalence signals hesitation and brooding. Boundary situations have a family resemblance to Wilfred Bion's (1989b) notion of "transitive thoughts" (p. 53). Bion states,

> We could regard artists, musicians, scientists, discoverers as those who have dared to entertain these transitive thoughts and ideas. It is in course of transit, in the course of changing from one position to another, that these people seem to be most vulnerable. (p. 53)

It is in the space of vulnerability that creative and intellectual work is done. When one founders between alternatives, one takes risks. The willingness to take risks is the mark of good scholarship. Working at the edge of understanding when uncertainty abounds, opens one to a certain amount of anxiety. Bion (1989b) remarks,

> At the same time they [those who take risks] are vulnerable to the observation of others who cannot tolerate the totality of the human personality, and therefore cannot tolerate someone who is so "mad," so "curious," or so "sane." (p. 53)

What "they" ("they" might be normalizing authorities, see Foucault for example) cannot tolerate is difference or alterity. Alterity is not merely an abstract word, it is a lived *experience*, it is an *experience* of Otherness.

Altered states of subjectivity might include the unspeakable: Beckettian states of The Mouth that opens but does not speak; The

Tramp who waits for Nothing; The Scream that is never heard (see for example Mary Aswell Doll's (1988)*Beckett and myth*). Witness Bergman's (1982) Ismael in *Fanny and Alexander*, who haunts Alexander. Robert Emmet Long (1994) suggests Ismael "whose biblical name gives the idea of estrangement and exile represents the destructive forces dwelling in imagination" (p. 170).

Bergman's (1960) tall strange figure who dresses in a black hooded robe and plays chess with a Knight in *The seventh seal* symbolizes the coming of the Knight's death, the knight symbolizes the Plague. The Plague gets everyone in the end.

Hallucinations perhaps—Dissociation—Deadness—Boredom—Anxiety—Melancholy—Pain—Dreams—Chaos—Depression—Dysfunction—Dystopia. Altered states are the unspeakable. Mary Aswell Doll (2004) tells us that when Beckett sent a manuscript to Routledge Press, the editor turned down his work because it gave him the "Jim Jams." Jim Jams are the unspeakable. Perhaps jim jams are too wild for some. What exactly are the jim jams? The unspeakable, that which is foreign—or Other—to the humdrum of everyday experience. Dan Zahavi (1999) argues that "one should be careful not to operate with too narrow a conception of the foreign. . . . Properly speaking it does not include only actually existing objects, but hallucinated and imagined objects as well"(p. 125).

Altered psychic states *alter* time and space. Cracked. Cracking up. Altered states may be *experienced*, fleetingly, slowly, simultaneously. Further, hallucinations and dreams that are, as Beckett (1965) might put it, "unnamables" shape the human condition. Michael Eigen (1996) suggests that other unnamable states might include "pockets of deadness" (p. 3) or "forms of disappearance other than dissociation and repression" (p. 43). Whether one would like to admit it or not, these states constitute part of the woof and web and fabric of human life. What do these states uncover or conceal? What do they crack open? Whatever these states reveal, their meaning depends on the use to which human beings put them.

Alienation and Otherness are not to be romanticized of course, but Otherness causes one to re-think one's place in the world in oppositional ways. Although alienation is emotionally and intellectually uncomfortable, it may open the way toward discovering Being, discovering altered states of subjectivity. Dan Zahavi (1999) comments that "Some phenomenologists . . . have claimed that the self-manifestation of subjectivity necessarily entails a self-alienation or self-transcendence, and that subjectivity only manifests itself to itself when it becomes Other to itself" (p. 112).

Thinking about different ways of experiencing Being—in the context of our scholarly and teacherly lives—is tremendously helpful, especially as educators in order to try to sort out what is going on emotionally both in the classroom and at the mystic writing pad. Michael Eigen writes about altered states that open up entire vistas usually ignored by psychologists and educators. Being "out of it," for example, is one such state. Eigen (1996) remarks,

> There is a kind of out of it, somewhere else dimension to our lives that needs acknowledgment. It takes many forms, from spacing out and going blank to vast depersonalization. We need empty, formless moments as respite and to reset ourselves. Blank immersion plays a generative role in creative processes. (p. 205)

When students stare out of the window teachers should leave them alone and let them be. Maybe they are "resetting" themselves, as Eigen suggests. It is important to be in touch with these kinds of altered states to try to figure out what one uses them to do.

Scholars who live on the edges of experience might benefit from the company of Other(s) who speak to a sense of alienation and alteredness/alterity. The problem of speaking about and through alterity and Otherness is that few are willing to grapple with these feelings in deep ways. In fact, Zahavi (1999) points out that it was Levinas who suggested that the philosophical tradition in the West has always had a problem grappling with Otherness. Zahavi (1999) states that

> According to Levinas, Western philosophy has been characterized by this attitude toward alterity. It has been inflicted with an insurmountable allergy, with a horror for the Other that remains Other, and has consequently and persistently tried to reduce alterity to sameness. (p. 196)

Comparative studies of difference become problematic for this very reason. If one attempts to cross the bridge of Otherness—by comparing one kind of otherness with another—one tends to collapse otherness onto sameness. Comparing differences across culture(s) levels the playing field and suggests that one difference is, at bottom, the same as another. But it is not. Otherness is singular, unique. This does not mean, though, that we cannot cross the bridge to talk. But crossing the bridge does not mean we "communicate" either. Do we ever really understand what it is we say to one another? On this score Lacan (1993a, 1993b, 1993c) is on the mark when he suggests that there can be no clear communication between people, if the Other is truly Other. There is always already a remainder (of Otherness) that gets misunderstood, that is non-translatable, that is the unspeakable.

Derrida repeatedly discusses Otherness throughout his work. Derrida (2002b) compares Otherness with the desert "but a certain desert, that which makes possible, opens, hollows or infinitizes the other" (p. 55). The desert is a place, a site of Otherness within. The desert of nothingness or emptiness or boredom plagues us all. But what to do when one arrives at the edge of the desert? One must make a choice. One might run toward water or one might find one's home being not-at-home in desertlike psychic spaces. Derrida suggests that we must welcome those who are not at-home and, in fact, live with the not at-home, Be not-at home. Become *hospitable* to those who experience Otherness. Derrida (2002b) points out,

> Hospitality is the deconstruction of the at-home, deconstruction is hospitality to the other, to the other than oneself, the other than "its other," to an other who is beyond any "its other." (p. 364)

Hospitality implies that a host honors a guest, especially when that guest does not feel at home. One does not live in a vacuum; lived-experience is a community affair. In a community of Others, though, communion may be impossible. The best we can do is to become hospitable. We may not be able to communicate with the Other, we may not be able to understand the other, but we could be civil to the Other. Civility is the principle upon which community might be based.

In order to Be not at-home, one needs a certain amount of company. Samuel Beckett (1980) remarks,

> A voice comes to one in the dark. Imagine. To one on his back in the dark. This he can tell by the pressure on his hind parts and by how the dark changes when he shuts his eyes and again when he opens them again. . . . You first saw the light on such and such a day and now you are on your back in the dark. A device perhaps from the incontrovertibility of the one to win credence for the other. That then is the proposition. To one on his back in the dark a voice tells of a past. With occasional allusion to a present and more rarely to a future as for example, You will end as you are now. And in another dark or in the same another devising it all for company. (pp. 7–8)

Scholars "devise it all" for company. We "devise" scholarly work so that we may have others listen to our cries and whispers, so that we may have company in our work. The life of the scholar is lonely because we work alone, and yet we are in the company of imaginary friends we find in books. Books sustain us in company.

When living in the desert of the university, one especially needs imaginary intellectual companions. These are the necessary delusions we keep in order to keep working within the site of the dystopic university. Scholars find illusionary companions in books. Company and companions. The solitary work of scholarship is not as solitary as we think. A room of one's own is necessarily peopled with the voices of others. Beckett (1980) says, "The voice comes to him now from one quarter and now from another. Now faint from afar and now a murmur in his ear. In the course of a single sentence it may change place and tone" (p. 15). Some illusionary voices, some delusional companions fill our heads because they are our company, our only real company when doing the work of thinking. The voices of our teachers, our parents, our analysts hover. What kinds of emotional and intellectual states do these companions evoke ? Texts upon which we draw fill the void, fill up the empty spaces, fill up dissociative states. Mary Aswell Doll (1988) talks of Beckett's work titled *Company* from which I take company while thinking about company and the work of intellectuals. She remarks that for Beckett,

> *Company,* Beckett's most child-centered fictional work, is narrated by an old man. Alone, the narrator lies in a dark room with no one to keep him company but the disembodied "one" of the narrative voice. (p. 76)

The narrative voice, for the scholar, is the voice that leads into the wilderness of scholarship and thought. It is a lonely journey, one that is otherworldly, yet grounded in one's own subjectivity and unconscious labyrinths. Sometimes the work demands a certain amount of distance from the everyday to get to the grist and meaning of everyday life. The scholar must travel deep within the place of no-where in order to get some-where. To find company, one must leave the world to find the world, one must shut one's self into the room with the computer and travel among texts in order to find others and ultimately self.

Thinking about the puzzling phenomenon of dissociation—a very strange altered state—Eigen (1993) asks "What is the subject when he is nowhere? What kind of nowhere is it?" (p. 112). Nowhere is the state between writing, where things are just not there, where nothing is in place. Interestingly, Wilfred Bion's writing, according to Gerard Bleandonu (2000), is remarkably dissociated. "At times Bion writes with an intense dissociation which interrupts the reader's associative links" (p. 48). One might be unconsciously drawn to Bion because one's own scholarship reflects the not at-homeness of dissociation. Beckett, like Bion is remarkably dissociative. Not surprisingly, Bion was Beckett's analyst (Knowlson, 1996).

Bion, more than any psychoanalytic thinker, captures a sense of being absently present or sometimes absent yet hovering. Is this madness? Perhaps a form of it. Eigen (1993) remarks "Given the ways madness works in our lives, how do we survive ourselves as long as we do?" (pp. 212–213) Some only barely survive. Eigen (1993) states that our childhoods are embedded within our psyches. For some, childhood was nothing but madness. There is no getting rid of childhood, even in adult life, especially if raised by crazy parents. The child is still there in one's psyche, the child within needs to be heard, needs a voice. Serge LeClaire argues that one must kill the child within in order to grow up and be an adult. The child within is nothing but a nuisance. Often times, this child interrupts adult life, especially if she hasn't been heard enough. Eigen (1993) draws poignantly on the work of Alice Miller. "Alice Miller has assembled documents that convincingly show how the physical and psychological cruelty of our childhood becomes the nightmare reality of adult existence" (p. 73). What Miller terms "poisonous pedagogy" (cited in Eigen, 1993, p. 74) becomes more apparent in later life. If the toxins of youth, via not good enough mothering, are not worked through psychologically, adult life crumbles. If life does crumble, perhaps one can trace this crumbling to childhood neglect. Otto Rank (1996) might suggest that one was not born well enough to survive the world. The trauma of birth is just too much for some, Rank would say. The birth trauma leaves one in a split-off state. Here, objects take on uncanny meaning; self and other are not clearly delineated; love and hate get mixed up. As Beckett puts it, "it's all symbiosis" (cited in Aswell Doll, 2004). The return of the repressed comes back tenfold in adult life. And so one founders in inbetween states, states without borders, as failed repression, pain and suffering overtake and take over a life.

If we are to teach children to understand their worlds, first they must understand themselves. What better way to understand the self than through study of the psyche, through the study of psychoanalytic ideas. Educators need to study deeply altered psychic states that cause feelings of Otherness.

Curriculum theorizing and psychoanalysis are natural bedfellows because both deal with the psyche and the world of the child. It is disturbing that psychoanalytic explorations, as they are related to educational processes, tend to be marginalized in the field of education. Derrida (2002b) comments that psychoanalytic writing and work are often overlooked (in the larger scene of the academy), or harshly criticized and ignored for one reason or another. Derrida's (2002b) most pointed remark is this.

> To ignore psychoanalysis can be done in a thousand ways, sometimes through extensive psychoanalytic knowledge that remains culturally dissociated. Psychoanalysis is ignored when it is not integrated into the most powerful discourses today on right, morality, politics, but also on science, philosophy, theology, etc. There are a thousand ways of avoiding such consistent integration, even in the institutional milieu of psychoanalysis. No doubt "psychoanalysis" . . . is receding in the West; it never broke out, never really crossed the borders of a part of "old Europe." (p. 89)

Psychoanalysis has certainly not been consistently integrated into education, even though Anna Freud and Melanie Klein's work have proved crucial for understanding children (for more on this topic see Britzman, 2003). Generally speaking, educational psychologists have little patience for psychoanalysis. Of course reasons for their dismissals vary. Some educational psychologists suggest that they have "moved beyond Freud." How can anyone move beyond Freud? Following Lacan's lead, Freud must be re-read; we must continually *return* to Freud. Freud is foundational for all kinds of psychology.

What is it about Freud that disturbs so many people? Why the continual controversy? Commentator Mortimer Ostow (1997b) suggests that psychoanalysis "is sometimes referred to as a Jewish science, as though there were something essentially Jewish about psychoanalysis" (p. 3). Of course there is nothing essentially Jewish about psychoanalysis. Freud's early followers, though, were all Jewish except Jung. Still, this doesn't make psychoanalysis essentially anything. Ostow points out that Jews might be drawn to psychoanalysis because being Jewish means being marginalized. Studying psychoanalysis helps one to better understand what that marginalization does to one's subjectivity.

Interestingly Ostow (1997b) states,

> Judaism and psychoanalysis share marginality: Jews are considered socially marginalized and psychoanalysis is considered academically marginal. Psychoanalysis offers an activity in which Jews struggling at the interface between the Jewish community and the non-Jewish world can express their conflicting needs. (p. 3)

Another commentator, Sarah Winter points out that Freud had his own suspicions about dismissals of psychoanalysis. Winter (1999) reminds us that Freud claims, in "The Resistance to Psychoanalysis," (henceforth RP 1925 [1924]) "[that] antisemitism lies behind the medical establishment's hostility to psychoanalysis" (RP, 222) (p. 36). Sander Gilman (1993) suggests, in his work on medical historiography, that the medical establishment has long been anti-Semitic. One begins to wonder about the "objectivity" of medical science. Gilman (1993) raises questions that plagued Freud during his lifetime such as these: If Jews are viewed as innately pathological, how then can they be doctors? How can they be psychoanalysts? Is psychoanalysis then a "pathological" science, if Jews invented it?

Educational psychologists, in particular, might dislike psychoanalysis for a variety of reasons. Perhaps they think it is hocus pocus and not scientific enough. Perhaps they do not like it because it is too literary. There are many reasons that people do not like psychoanalysis. Why not just prescribe prozac? That is the American way. Forget the analysis, give the kid a pill. Your kid is nervous? ADHD? Take a pill. There is even Prozac for dogs!! Is your puppy anxious? Give him prozac. How about that! Forget about the painstaking work of analysis. Americans are too much in a hurry and would rather spend money on Hummers than have themselves or their children analyzed.

Schools and Madness

"Why do you shake hands?" "Because I can't eat my students."

(Cited in Bleuler, 1911/1950, p. 81)

At this juncture let us begin unpacking the notion of madness. This inquiry attempts to ask the following questions: how does it *feel* to be mad? Why is it important to understand what it feels like to be mad? What do children and schooling have to do with madness? What if some children border on madness? Are questions such as these relevant to the study of education? Some would think not. What does it feel like to live in a state of Otherness, a site where children might founder?

In the educational literature on Otherness, scholars do not take the plunge into extreme states of Otherness. They give lip service to Otherness. Embrace difference, they say. Well, how can you embrace difference if you don't even know what it is? Madness is, I admit, an extreme example of Otherness, but perhaps we should begin with the extreme.

Madness may be genetic or environmental. Perhaps it is a mixture of both. And maybe asking about causes are not really important. Focusing an inquiry on causes gets off the point. The interest in this study is in deconstructing mad states of being, extreme states of Otherness, so that educators can get beyond slogans like "embrace difference." We must get beyond slogans to understand emotionally what difference or Otherness feels like. Otherwise, we do not have insight about the Other.

Let us dwell for a moment on William F. Pinar's (1975/2000) groundbreaking paper titled "Sanity, Madness and the School." Pinar (1975/2000) eloquently states thus:

> One theme common to almost all [school] criticism is the contention that the schooling experience is a dehumanizing one. Whatever native intelligence, resourcefulness, indeed, whatever goodness is inherent in man deteriorates under the impact of the school. The result is the one-dimensional man, the anomic man, dehumanized and, for some critics, maddened. (p. 359)

Pinar's (1975/2000) essay forcefully expresses one of the most serious problems of public education that plagues the American landscape still. School drives creative people into the ground; school unravels children's native intelligence. Of course, students and teachers are always already troubled/maddened by the larger corporate American culture even before stepping inside the schoolhouse or university. But stepping inside the schoolhouse only worsens one's already vulnerable emotional and intellectual state of Being. American culture is conformist and conservative, no doubt. American schools do not tolerate difference, period.

It is time to take a leap into varieties of madness that plague our school children and teachers. Not that all school children are insane! But many young people are not exactly healthy either. School can make kids ill, indeed. Like the not good enough mother, the not good enough schoolhouse shuts kids' emotions down and drives them into schizoid states. Pinar (1975/2000) declares,

> When one is "absent" for much of a six hour period, day after day, year after year, one becomes "absent" most of the day, day after day. One is not in the "real world." In fact, one may be designated, at some point, as psychotic. On the other hand, when one is present, most of the time, and that "presence" is achieved by violence, e.g. Paul forced his

> daydreams from his head, rendering his fantasy life lifeless, one loses an integral part of himself. (p. 363)

If school does not allow for phantasy and play, children suffer. Melanie Klein (1955a/1993) points out that the inability to phantasize and play are signs that children are fixated in what she terms the paranoid-schizoid position. When children are severely inhibited, withdrawn, and obviously absent, something is wrong. Thus, I argue, as did Pinar 20 years ago, that the not good enough schoolhouse is partly to blame for children's' ill health. Scholars dwelling in universities do not fare much better because universities too are places where phantasy and play are not allowed, where creativity is questioned, especially if it is not tied directly to the market (Readings, 1996). And yet, good scholarship and good teaching have much to do with the ability to both phantasize and play, to explore, creatively and deeply, one's subjectivity and one's place within the academy. Phantasy and play allow educators to founder, hesitate, feel ambivalent, dream, embrace oddity. Deborah Britzman (1998) contends that education might nurture "proliferating identities" (p. 219). But schooling shuts identities down. Institutions—such as schools and universities—inculcate sameness, orthodoxy.

What is to be done? I take my lead from Mary Aswell Doll (1995) who argues that scholars need to "turn the curriculum inside out" (p. 63). How does one go about doing this? Scholars have commented again and again on the wasteland of the school and university as sites of normalization and control. To turn the curriculum inside out one might also take Michael Eigen's (2001a) lead. As I previously mentioned, he suggests that one must "make room" (p. 5) for psychic states like pain, suffering, anxiety, dissociation. Eigen (2001a) warns, however, suffering cannot be "cured." "Making room" for pain does not mean fixing it. Eigen (2001a) states,

> Of course, one does not "cure" rupture and despair, indelible as they are. But it is possible to enter the living stream of a relationship that grows with rupture—return, that makes room for despair. (p. 118)

Scholars must "make room" for madness in the curriculum—in order to turn it inside out—since madness is always already there. That madness is a taboo topic in the field of education puzzles. School is such a troubling site. Why are kids turning to guns? Why are kids unable to think? Why are kids bored and angry? Sitting in our classrooms are pained and depressed children. Since Columbine, scholars must begin to think about why children are so depressed, apathetic, and smug. By the time our children reach college many are already emotionally ruined. They are ruined for a variety of reasons, but mostly because of adults' indifference and inability to reach them. Adding more standardized tests to the curriculum only worsens matters.

Moreover, children are damaged by adults' inability to grapple with Otherness. In fact, Philip Wexler (1996) comments that in the field of education the concept of "difference" is not adequately dealt with. Difference is a trope that has become ineffective and perhaps meaningless because, in reality, teachers do not want to begin to think about the unthinkable, the unimaginable. Wexler (1996) states,

> But, this difference, whether as "otherness" or as the transcendental direction of what is "entirely different," or "otherwise," remains empty, void of substantive definition except as a virtuous principle, or as an approval of conventional pluralism. The burning question of how to see things clearly and how to live the present differently is not answered, unless the character of the difference can be elaborated. (p. 4)

Elaborating the trope of madness—as an extreme example of difference—might not clarify lived experience, but it might deepen one's understanding of how difficult it is to understand what it feels like to live on the other side of the moon. And there are children who do indeed live on the other side of the moon and sit in our classrooms. In fact, Melanie Klein (1930/1992) alarmingly states "I have become convinced that schizophrenia is much commoner in childhood than is usually supposed" (p. 230). Not that all children are schizophrenic, of course. But many children do, in fact, suffer various emotional traumas that educators simply ignore or do not think about.

Teachers, professors, and students alike suffer various emotional upheavals at some time during their lives. What teachers and professors ignore is the fact that emotional upheavals might founder somewhere between sanity and madness. In fact, many psychoanalysts argue that it is quite difficult to tell the difference between sanity and insanity (Klein, 1952/1993; Bion, 1976/1993b; Riviere, 1991; Gilman, 1994; Saas, 1992). Melanie Klein (1957/1993) suggests that "a residue of paranoid and schizoid feelings and mechanisms, often split off from the other parts of the self, [and] exists even in normal people" (p. 210). Stephen Mitchell and Margaret Black (1995) remark that "For Klein, the psyche, not just of small children but of the adult as well, remains always unstable, fluid, constantly fending off psychotic anxieties" (p. 87). According to James Grotstein (1993),

Bion argued that "most of us have a neurotic personality and a psychotic personality" (pp. 11–12). Some analysts note that psychotic and non-pyschotic parts of the self live side by side. Michael Sinason (1999) says that "Jenkins (1999) shows how the existence of a 'psychotic self' cohabitating with a 'non-psychotic self' can be seen in many different psychopathologies" (pp. 51–52). These altered states of Being might alarm. One does not like to think of oneself as tipping over the edge. Cracking up. Even if one recognizes one's own capacity for experiencing altered states, one tends to dismiss this capacity of cracking. Some reduce psychic upheavals to having a bad day, or "losing it." But what does one mean when one says I've lost it? Has one lost one's sanity? Has one lost one's bodily ego? If it is lost where does it go? What does one mean when one says I've had a bad day? What does that badness entail? What does one mean by being bad? Is having a bad day more than losing one's rhythm?

How might one dare to think like this against the backdrop of American conformism? How could teachers and professors dare to think about their own psychic transformations when so mired in the busy world of school life? Teachers must dare to think about Otherness. And this is why William Pinar (2004) has insisted over and over again that one must think autobiographically. What does it mean to think about one's own psychic mechanisms? These are complex questions that demand complex theoretical, analytic work. Who are we talking to when we teach? Who are these students anyway? Where are they? Are students in our company or are they zoned out? Where ever they are—they are elsewhere—they are completely and utterly Other to us. Zahavi (1999) teaches that for Levinas, the Other is Other. We may never be able to fully articulate that kind of Otherness. Zahavi (1999) writes,

> According to Levinas, a true encounter with the Other, is an experience of something that cannot be conceptualized or categorized. It is a relation with a total and absolute alterity. (p. 196)

Grappling intelligently with forms of alterity such as madness means that one cannot tidy up what these tropes might mean or how altered states might be experienced from a phenomenological or psychoanalytic point of view. Experiencing Otherness, throws us, as Michael Eigen (1993) suggests, "to the edge" (p. 1). The edge might be a place where children remember abuse or a place where they were emotionally wounded by their parents. And since, as Eigen (1996) aptly puts it, "one cannot make savage wounds go away" (p. 21), educators might take a look at these savage wounds and "make room" for them in the curriculum. If curriculum is the complicated conversation between teachers and students (Pinar, 1995), one must examine wounded states of Being. Being wounded means being complicated. Our students are complicated—and sometimes wounded—beings. The institutions in which children spend much of their lives—schools and universities—do not tend to their wounds. Schooling only deepens the wounds.

Mad States of Being

Let us now explore psychoanalytically and phenomenologically mad states. Let us listen to Derrida's (2002d) suggestion so that we may be ready for the coming-of-the-Other. He argues that one must, "Let oneself be swept away by the coming of the wholly other, the absolutely unforeseeable [inanticipatable] stranger, the uninvited visitor" (p. 361). Readers might suspend rational judgment and try to be "swept away" by descriptions of mad states. The rational mind might struggle to make sense of the non-rational. Readers must just go with it—as it were—in order to get some understanding of what it feels like to be totally Other. Emmanuel Levinas (1998) suggests that it is through language that one learns about the Other, that one comes face to face with alterity. Levinas contends that "Language, as the manifestation of a reason, awakens in me and in the other what we have in common. But it assumes, in its expressive intention, our alterity" (pp. 25–26). As Mary Aswell Doll (2000) claims, "Language, as James Joyce (1939) might pun, is not just the letter but the litter—the leftover that lingers somewhere in another mindplace, and beckons" (p. xv). Otherness, alterity, difference, and madness might be found in these litters, these leftovers, these left-out-of-the-canon taboos that do not beckon a reason, but a non-reason. It is in the non-reason that sense cannot be made, yet it is here that we might make sense of seeming non-sense. Upon approaching the language of the mad, one note of caution is in order. Levinas (1998) warns that the language of the Other, must never be whitewashed, never trivialized or cleaned up. Levinas (1998) states,

> The interhuman is also in the recourse that people have to one another for help, before the astonishing alterity of the other has been banalized or dimmed down to a simple exchange of courtesies that has been established as an "interpersonal commerce" of customs. (p. 101)

Films like *A Beautiful Mind*, hollywoodify madness. *A Beautiful Mind* made it seem almost fun to be insane. Being insane is no fun, as psychoanalysis can attest. And why the title? There is nothing "beautiful" about insanity. We must never glorify mad states of being, which is what the title of this film does. Madness must be re-presented without glossing over it, dimming it down, or dumbing it down.

American ego psychologists have been accused of turning the language of madness into niceties. Russell Jacoby (1997) remarks that Alfred Adler was guilty of "retreat[ing] to pleasantries" (p. 19) in order to "assuage the pain of the familiar" (pp. 44–45) and to sell out to "common sense" (p. 19). Jacoby (1987) states that these "liberal revisions" (p.19) were exactly out of step with Freud. "Orthodox psychoanalysis is oriented in the reverse direction: toward uncommon sense, exactly the farfetched" (p. 20). But it seems that when talking of the unspeakable—like madness—many prefer common sense, pleasantries, and "civilized" language. Wilfred Bion (2000) remarks that people do not like to think "otherwise." He states that "Libraries from the beginning of time have also been burned down because they are such a terrible irritant; people hate having their thoughts stirred up" (p. 134). As a matter of fact, Bion (2000) states, "It is dangerous to consider anything" (p. 214). Jacques Lacan (1993d) intuits that "we are afraid that we'll go a little bit mad as soon as we don't say exactly the same thing as everybody else" (p. 201). But how can one talk about "a psyche undoing itself" (Eigen, 1996, p. 14) in pleasantries? As Mary Aswell Doll (2000) states, "The shock of confrontation with that which is utterly Other helps push dogma off its stone" (p. xv). Let us confront head on—and perhaps shock—ourselves—world(s) of the mad.

It is not as if the Western cultural tradition has not known and written about madness. It is not as if being mad is anything new at all. In fact, deeply rooted in the Jewish tradition, Ezekiel, the mad prophet, has historically stirred up biblical commentators. Indeed, Abraham Heschel (1969) suggests that the most shocking people, the most "disturbing people who ever lived" were the Hebrew prophets (p. ix). Ezekiel was mad, schizophrenic, psychotic—use whatever name you like—he was mad (see E. C. Broome, cited in Bloch, 1997). Ezekiel cannot be reduced to pleasantries. He pulled out his hair, ate dung and scrolls, laid on his side for days at a time as rigid as a corpse, hallucinated. Michael Sinason (1999) comments,

> Hair pulling is part of the historic imagery of madness and madhouses through the centuries. The best-known works of art illustrating this are probably Caius Gabriel Cibber's two huge stone figures of madmen named "Raving and Melancholy madness," which surmounted the

> gates of the New Bethlehem Hospital built in Moorfield's in 1675, following the great fire of London. (p. 46)

Bleuler (1911/1950) remarks that "some patients tear their hair out by the roots, frequently in a very definite pattern" (p. 186). In Ezekiel 5:1 we read the following bizarre passage:

> And you O Mortal, take a sharp sword; use it as a barber's razor and run it over your head and your beard; then take balances for weighing, and divide the hair. One third of the hair you shall burn in the fire inside the city. (NRSV, 1989, p. 794)

Along with hair pulling, catatonic psychotic patients—such as Ezekiel—are so rigid that they can be lifted up as if they were statues, they can be moved while remaining in a rigid state. Bleuler (1911/1950) comments,

> Indeed it is not at all rare to meet with a patient who can assume and maintain a certain position for months at a time; he will seem quite rigid too, when one attempts to move his limbs . . . one can move the patient's whole body as if it were a piece of wood. (p. 180)

Bleuler's description of catatonia could describe Ezekiel's behavior in 4:4:

> Then lie on your left side, and place the punishment of the house of Israel upon it; you shall bear their punishment for the number of days, three hundred ninety days, equal to the number of the years of their punishment; and so you shall bear the punishment of the house of Israel. When you have completed these, you shall lie down a second time, but on your right side, and bear the punishment of the house of Judah; forty days. (1989, p. 793)

Psychotics say that voices tell them what to do (in Ezekiel's case he hears the voice of God). Psychotics suffer from omnipotence and an overly strong super-ego (Ezekiel mentions the word punishment at least five times in this passage alone). Psychotics, Melanie Klein (1929a/1992) comments, suffer from,

> a displaced relation to reality . . . the wish-fulfillment is negative and extremely cruel types are impersonated in play . . . the ascendancy of a terrifying super-ego which has been introjected in the earliest stages of ego-development. (p. 207)

The cruel "God" forces Ezekiel to perform bizarre acts; this "God," one might assume, is little more than Ezekiel's introjected, not good enough mother, and a constitutionally disturbed self. No pleasantries here. What is amazing, though, and perhaps not surprising, is that conservative biblical scholars like Bloch deny that Ezekiel was, in fact, mad. Bloch *unconvincingly* claims that "The psychoanalytic approach [to Ezekiel] has been rejected by commentators and psychiatrists alike" (p. 10). What psychiatrist would deny these remarkably psychotic passages? But then contradictorily Bloch (1997) says,

> E.C. Broome concluded that Ezekiel was a true psychotic, capable of great religious insight but exhibiting a series of diagnostic characteristics: catatonia, narcissistic-masochistic conflict, schizophrenic withdrawal, delusions of grandeur and of persecution. In short, he suffered from a paranoid condition common in many great spiritual leaders. (p. 10)

I do not discount the insight madness brings. I agree with R. D. Laing (1960), Carl Jung (1963), and Joseph Campbell (1972) that madness can yield insight. However, what I do discount is Bloch's refusal to admit that the great Hebrew prophet was a madman!!

Does this make his prophecy any less important? I do not think so. R. D. Laing (1960), commenting on Ezekiel, remarks that "the cracked mind of the schizophrenic may let in light which does not enter the intact minds of many sane people whose minds are closed" (p. 28). Eugene Bleuler (1911/1950) comments that psychotic breaks help artists.

> However, we also know that several very well-known artists and poets (e.g. Schumann, Scheffel, Lenz, van Gough) were schizophrenics. It cannot be ruled out that very mild forms of schizophrenia may be rather favorable to artistic production. The subordination of all thought-associations to one complex, the inclination to novel, unusual range of ideas, the indifference to tradition, the lack of restraint, must all be favorable influences. (pp. 89–90)

Although psychosis might open one to creative forms of thought, one must be cautious not to romanticize schizoid states. Deleuze and Guattari (1987, 2000) tend to romanticize schizoid states in their work on what they term "schizoanalysis." This troubles. Although their work on schizoanalysis is interesting, it does tend to trivialize. "Schizoanalysis is the art of the new" (1987, p. 203), argue Deleuze and Guattari. Studying schizophrenics, Deleuze and Guattari (1987) suggest, helps us find "lines of flight" to "break through" (2000,

p. 277) rigid spaces that oppress. The schizophrenic somehow "scrambles all the codes" (2000, p. 15) and makes a mockery or "shambles" (2000, p. 135) of Freud and the entire psychoanalytic and psychiatric tradition. Deleuze and Guattari suggest that neither the psychoanalytic nor psychiatric tradition has bothered listening to schizophrenics. Perhaps there is some truth to this. Psychoanalysts, especially those of Freudian and Kleinian bents, have been accused by Deleuze and Guattari of foreclosing their interpretations too soon, even before the psychotic enters the room. A scathing critique of Freud's well-known child analysis of "Little Hans," who suffered from phobias, is offered by Deleuze and Guattari (1987).

> Look at what happened to Little Hans, as example of child psychoanalysis at its purest: they kept on BREAKING HIS RHIZOME and BLOCKING HIS MAP, setting it straight for him, blocking his every way out, until he began to desire his own shame and guilt in him, PHOBIA, (they barred him from the rhizome of the building, then from the rhizome of the street, they rooted him in his parents' bed, they radicled him to his own body, they fixated him on Professor Freud. (p. 141)

Reductionistic interpretations of psychic states, of course, are problematic. Some analysts might be reductionistic in their interpretations. There are, of course, poor analysts. But there are also good ones who do not reduce the complex to the simple. Deleuze and Guattari assume all analysts are poor. This is an unconvincing argument on many counts. Good analysts—or I should say in the Winnicottian fashion "good enough" analysts—are those who use psychoanalytic language to help people understand who they are, not block their "rhizomes" in advance. If anything, Adam Phillips (2001) and Michael Eigen (1993) point out, psychoanalytic language should be used to help us understand how utterly Other we are. Language can, though, be used in a naive and reductionistic way. This is indeed what Wilfred Bion fought against. Bion was disturbed when analysts used Kleinian interpretations to block the rhizomes of difference in advance. Mitchell and Black (1995) comment that "Bion became dissatisfied with the formulistic way many clinicians applied psychoanalytic concepts (including Kleinian concepts) and took a particular interest in trying to explore and convey the dense texture of ultimate elusiveness of experience" (p. 102). Like Bion, Eigen and Phillips attempt to explore density and elusiveness throughout their work. I do not get a sense that either one of these analysts is reductive in his thinking. In fact, they continually explode taken-for-granted notions of self.

It has been documented, though, that psychoanalytic institutes, especially because of standardization, encourage technical and predictable interpretative work. Donna Bentolila (2000) remarks that "Kernberg sharply critiques the psychoanalytic institutes, which, in his opinion, have been transformed into technical schools" (p. 323). Educationists are well aware of the problematics of standardized curricula. Still, there is much in the literature that suggests that some analysts subvert standardization practices which only serve to normalize. Certainly those analysts associated with the New York Institute of Psychoanalysis (like Michael Eigen and Jessica Benjamin) are doing groundbreaking work and have somehow managed not to allow themselves to become slaves of standardization.

Standardization in psychoanalytic institutes mirror the ongoing tragedy of the push toward standardization in schools and universities in the United States. Like the maverick analysts I mention above, curriculum theorists (see, e.g., Aswell Doll, 2000; Morris, 2001; Morris and Weaver, 2002; Britzman, 2003; Pinar, 2004) work toward deeply exploring lived experience in schools and universities that is *not standardized*. How can one standardize lived experience? Should not educational research study the ways in which we live within the sites of schools? To understand youth, one must understand the culture in which youth live. Youth culture(s) are highly complex and diverse. No standardized curriculum can serve the needs of a diverse and complex body of students. Curriculum theorists have long struggled against the standardized testing movement. Education should be about the possibilities of developing idiosyncratic knowledge(s), eccentric ideas. Educational research should study our complicated relations with one another in an uncertain and chaotic world. How can one standardize a post–9-11 world? How can one standardize a Columbine? Standardized tests do not help our children understand the world in which they live. Standardized tests damage children's sense of self-worth. Standardized tests wound children's psyches forever. Americans will continue to be miseducated and undereducated as long as people continue to stand for standardization.

Deleuze and Guattari's (2000) contention in *Anti-Oedipus* is that psychoanalytic work reduces every issue to the Oedipus complex. Certainly the Oedipus complex is central to analytic work but there is more to it than that. For example, Otto Rank (1996), in 1924, began to take a contrary position to Freud by arguing that the pre-Oedipal years were more crucial than Oedipal or post-Oedipal years. In fact, Rank was a forerunner of the object-relations movement (Kramer, 1996). The trauma of birth—Rank's contribution to psychoanalysis—had everything to do with the child's relation to the mother. Other maverick analysts like Georg Groddeck (1961), focus not on Oedipus but on the unconscious. In a letter to Freud dated 1917, Groddeck discusses his notion of the "It." The "It" has little to do with the Oedipus complex. For Groddeck, the "It" drives us to do things. The "It" is a mystical creature that moves our lives forward. The notion of the "It" is reminiscent of Freud's Id, but Groddeck's "It" is mystical, whereas Freud's Id is scientific—or so he thought. Likewise, Melanie Klein's (1930/1992; 1946/1993) work is *not* grounded in the Oedipus complex, it is grounded in object-relations which are pre-Oedipal.

Deleuze and Guattari, thus, exaggerate their claim that all psychoanalysis is about King Oedipus. They make no attempt to qualify their statements or nuance their pronouncements. In an effort to debunk all of psychoanalysis they suggest that Oedipus is dead. That is, they suggests that the Oedipus complex is not useful when examining why it is that people become psychotic. But other analysts have been saying this for years, (see, e.g., Klein, 1946/1993; Jung, 1963; Rank, 1996) while not totally dismissing the usefulness of psychoanalytic work.

The reason Deleuze and Guattari perform a "schizoanalysis" is to point out that not all people's craziness can be reduced to the Oedipus complex. In other words, Deleuze and Guattari attempt to show that the psychotic cannot be Oedipalized in fact—the psychotic cannot be analyzed at all. Rather, "Schizoanalysis proposes to reach those regions of the orphan unconscious—indeed 'beyond all law'—when the problem of Oedipus can no longer even be raised" (pp. 81–82). What troubles here is that Deleuze and Guattari (2000) offer a scathing *Destruction* rather than a deconstruction of psychoanalysis—without putting any useful in its place. Schizoanalysis is not constructive because it does not generate ideas that might suggest alternative treatments for psychotics. "Destroy, destroy, The task of schizoanalysis goes by way of destruction—a whole scouring of the unconscious, a complete curettage. Destroy Oedipus, the illusion of the ego, the puppet of the superego, guilt, the law, castration" (Deleuze and Guattari, 2000 p. 311). Okay. Then what? Whither Oedipus? Deleuze and Guattari are exactly wrong here. "The law of the father," "castration complexes," and so forth, are useful metaphors when analysts put them to work in ways that help patients understand troubling aspects of self. These metaphors help articulate what troubles. Of course few analysts would reduce everything to x, or y, or z. But analysts must have some way of talking about psychoses.

Deleuze and Guattari (2000) cynically ask psychoanalysts why they should help psychotics in the first place. "Why try to bring him back to what he escaped from" (p. 23). The psychotic's response to analysis—according to Deleuze and Guattari is "they're fucking me over again" (p. 23). The psychotic may "love" his delusions, as Lacan (1993b, p. 157) suggests, but maybe he or she does not love living in terror. Michael Eigen (1993) explains that "For the psychotic individual, the natural interweaving of self and other may turn into a terrifying sense of dissolution or invasion" (pp. 32–33). Terrifying. Is it ethical for analysts to give up on psychotics? Eigen (1993) poetically explains Bion's take on psychotic states:

> The nameless sense of catastrophe Bion points to moves between nothingness—somethingness—everythingness. It threatens to hurl the personality into an unimaginable abyss beyond oblivion, a horrific spacelessness in which there is no direction or valence other than horror itself. It fuses chaos and nothingness, scattered noise and blankness. (p. 119)

Terrifying, horrific, catastrophic. These are not feelings or states that should be made light of, banalized, trivialized, whitewashed, or romanticized. And certainly the psychoanalytic community should never abandon psychotic patients. With the help of anti-psychotic medicine* and talk therapy some patients may benefit from analysis. Others perhaps not. But still this does not mean that analysts should simply throw up their hands. And this is exactly what Deleuze and Guattari end up doing. Deleuze and Guattari claim that analysts do not listen to schizophrenics—anyway—so what's the point. However, analysts like Klein and Jung did pay attention to psychotics. Others like Bion worked with psychotics and many contemporary analysts of the Jungian, Kleinian, and Bionic bent continue to work with psychotic patients. Deleuze and Guattari (1987) seem even to mock psychosis as they state

> The body without organs [a reference to Paul Schreber] is not a dead body but a living body all the more alive and teaming once it has blown apart the organism and its organization. Lice hopping on the beach. Skin colonies. The full body without organs is a body populated by multiplicities. (p. 30)

* Anti-psychotic medicine has taught us that psychosis is partly a result of chemistry and biology but not reducible to either.

Paul Schreber and many other psychotics do dwell on organs, on bodies "without organs," on the mutilation of organs, on hallucinations about organs. But these are not states to be mocked. Who was the man Schreber? Who was the man who complained about organs? Artaud (1988f) said, "One must experience the real unraveled void, the void that no longer has an organ" (p. 72). It troubles to go into the void, but it is necessary for analysts to try to understand what this is. In what specific ways did Artaud suffer? Should not analysts want to find this out? Artaud's memoirs are a great pedagogical tool for learning about psychotic states. He is a must read. Artaud was clearly tormented. There is nothing to mock or trivialize about feeling states of torment. I find that Deleuze and Guattari are simply too smug and flippant around these issues. In fact, I find their work on schizoanalysis repulsive. Artaud (1988c) states "I suffer from a horrible sickness of mind" (p. 31). What is the alternative to helping someone like Artaud, abandonment? That is what Deleuze and Guattari suggest. Abandoning patients is simply unethical. Medically, doctors and lay analysts alike have a responsibility to take care of people like Artaud. Artaud expresses a terrifying experience in a letter to George Soulie De Morant. Artaud (1988b) complains that he is

> at the mercy of a kind of terrifying crushing and tearing of consciousness, truly baffled with respect to my most elementary perceptions, unable to connect anything, to assemble anything in my mind or still less to express anything, since nothing could be retained. (p. 288)

Do Deleuze and Guattari take Artaud seriously or do they simply reduce him to "lice hopping on the beach?" Do they take psychosis seriously? In fact, they mock it. Artaud (1988e) remarks: "For a madman is also a man whom society did not want to hear and whom it wanted to prevent from uttering certain intolerable truths" (p. 485). An interpretive move that cuts off madness by not listening seriously or by mocking is not responsible. In fact it is reprehensible. We lock people away in mental institutions because we don't want to deal with people who are out-there, who are, in a word, ill. "Civilized" society wants nothing to do with the insane. It is always somebody else's problem. Let us make it our problem, let us try to understand. If we—as a society—do not want to deal with mental issues—whatever other issues do we really—at the end of the day—not want to deal with? The fact of the matter is, people do not like difference, they do not like what they do not understand. If we can, for a moment, try to understand this extreme form of alterity, perhaps we can begin to understand other forms—not so extreme—and begin to become a more ethically responsible people(s). Otherwise, we resort to barbarism. Locking people up is, in fact, barbaric. Shunning the Other is criminal.

Reading the Mad: Interpretive Openness

As I have previously (Morris, 2001) grappled with the problematic of reductionistic readings of psychoanalysis, I will not belabor the point here. A psychoanalytic/phenomenological hermeneutic is one that is open-ended, always already embracing strangeness. As Adam Phillips (2001) suggests, the point of doing interpretive work with patients is "to make the patient the good-enough poet of his own life" (p. 9). The good enough analyst encourages "the possibility of an eccentric life" (p. xiii). A good enough psychoanalytic hermeneut might take advice from Deborah Britzman (1988) who argues that educators, like analysts, might begin "accounting for the relations between a thought and what it cannot think" (p. 212). Thinking unthinkable thoughts opens one to uncomfortable states. When dealing with such painful subjects as psychosis, one must be willing to experience dis-comfort. Thomas Ogden (1989), in his book titled *The primitive edge of experience*, remarks,

> A reader, like an analysand, dares to experience the disturbing feeling of not knowing each time he begins reading a new piece of writing. We regularly create the soothing illusion for ourselves that we have nothing to lose from the experience of reading, and that we can only gain from it. This rationalization is superficial salve for the wound that we are about to open in the process of our effort to learn. (p. 2)

Dennis Sumara (1996) comments that "the practice of reading in one's life . . . means being prepared to have the order of one's life rearranged" (p. 9). Indeed. Sumara argues that reading "alters us phenomenologically" (p. 108). Beautifully stated. Frightening. Sumara asks "Who will this reading ask us to be?" (p. 152) Well, certainly studying psychosis asks us to take the plunge into strange waters. Some would rather not dive in. Why would that be—from a psychoanalytic perspective. Is it because their personae are too well defended? Michael Eigen (1993) comments that

> the masks of sanity people wear do not always work well for them. The addiction to elements of personality one considers sane can be destructive. Some people would rather die than risk madness. (p. 339)

But if one wishes to take this journey and delve into psychotic states, one has to risk one's psychic foundation; one must begin foundering and go with it. Eigen (1993) suggests when he works with psychotic patients that

> we must allow ourselves simultaneously to enter, yet maintain distance from the phenomena we encounter. A certain slippage between one level of discourse and another must be tolerated. Descriptions of psychotic experience, structures, and dynamic operations slide into each other. (p. 31)

Both delving and distance are indeed necessary when dealing with difficult emotional material. An intuitive opening toward Being only emerges at the edge of experience and at the edge of what Christopher Bollas (1989) terms "the unthought known" (p. 18). Interpreting difficult texts—like interpreting different life experiences—creates the space for readers to not know, to not anticipate what is to come and to be open to not anticipating. To not judge in advance or dismiss upfront what might appear—becomes key. Distance, delving, not knowing, and not judging in advance—might constitute what Bollas (1989) has in mind when he talks of gathering together "a set of interpretive references" (p. 63)—when analyzing a patient or—I would add—a text. These interpretive references are found in the *via negativa*. This is the space of no-where. The *via negativa* is a Beckettian state of "blathering in the void" (Aswell Doll, 2004). In order to gather together this set of "interpretive references" about blathering in the void, one might consult Freud or Bion. Nina Coltart (2000) explains,

> We have been waiting attentively, in Freud's own words, "for the pattern to emerge." Those of us who were fortunate enough to be taught by the late Dr. Bion value the stress which he laid on the need to develop the ability to tolerate not knowing; the capacity to sit it out with a patient, often for long periods, without any real precision as to where we are, relying on our regular tools and our faith in the process to carry us through the obfuscating darkness of resistance, complex defences, and the sheer unconsciousness of the unconscious. (p. 3)

Waiting it out for patterns to emerge, not knowing precisely where these patterns will take us is key to doing good textual hermeneutic work as well. Part of the not knowing is, of course, related to the unconscious. Acknowledging the "sheer unconsciousness of the unconscious" drives one to interpret with more freedom and agility.

Interpretative work that values open-ended, patient waiting "makes room" (Eigen, 2001a, p. 5) for nuance and uncertainty, ambivalence and foundering. Perhaps the text will speak or perhaps the text will not speak at all. Adam Phillips (1993) reminds us that if our focus of interpretation or our site of interpretation is ultimately unconscious, we simply cannot know in advance what we are doing. Phillips (1993) says,

> If there is an unconscious, what is the analyst [or hermeneut] doing when he thinks he knows what he is doing? Each analysis improvises within mostly unconscious theoretical constraints. (p. xiii)

The improvisation of theoretical analysis, though, is not without structure or study. It is not, on the other hand, prefabricated in advance; theoretical work is what it is through the sheer grit of thinking and writing, being and feeling, studying and living deeply. There is no formula for doing theoretical work in advance, there is no narrative outline, there is no plot to follow. Nina Coltart (2000), like Adam Phillips, talks about the difficulties of not knowing where one is going when one is doing analytic work. Coltart (2000) states these difficulties eloquently:

> It is of the essence of our impossible profession that in a very singular way we do not know what we are doing. Do not be distracted by random associations to this idea. I am not undermining our deep, exacting training; nor discounting the ways in which—unlike many people who master a subject and then do it, or teach it—we have to keep at ourselves, our literature and our clinical crosstalk with colleagues. All these daily operations are the efficient, skillful and thinkable tools with which we constantly approach the heart of our work, which is a mystery. (p. 2)

Coltart's comments are relevant to the work of curriculum scholarship. Our work as curriculum scholars is a mystery too. We have our tools, our books, our discussions with colleagues, but the ways in which we go about theorizing remains ineffable. Lived experience is anything but straightforward. Heidegger (1927/1962) suggested that after a long period of dwelling in thought, one might enter a "clearing." Exactly what "the clearing" entails I do not know. But entering a "clearing" suggests—generally speaking—that one is better able to make sense of what was before confused and tangled up. When we are talking, however, about psychosis, there is no clearing. All we have is a mess. No matter how much thought you give to psychosis it is still a most unclear subject. There is no "clearing" in psychosis. Moreover, Michael Eigen (2001b) suggests that one should not try to clear up the mess. He states,

> One never undoes the tangles of real living. I am not sure that undoing knots is a good model for many problems we endure. More importantly is appreciative access to density, to navels of experience. Wave after wave of experience passes, if one notices, and one cannot with assurance tease nourishment from toxins. (p. xx)

It is interesting that phenomenological and psychological "density" gets expressed, one way or another, in the work of curriculum studies, poststructuralism, deconstruction, and psychoanalysis (see, e.g., Derrida, 1987, 1992, 1998a, 1998b, 2001; Bion, 1991a, 1991b; Lacan, 1993a, 1993b, 1993c, 1993d; Aswell Doll, 2000; Phillips, 2001; Jardine, 2002; Eigen, 2004; Pinar 2004). Thus, the aim in this chapter is not to untangle psychic states. The experiences examined here are dense, complex, and that is the way they will stay. This study does not try to clarify what cannot be made clear.

This approach to the subject of madness is variegated. A variety of psychoanalytic thinkers are consulted so as not to limit this perspective to any one school. I do not want to limit this study by focusing, say, only on Klein's or Jung's explanations of psychosis. A wide array of analysts will be consulted. Further this is an interdisciplinary approach to the subject at hand. I am not an analyst, nor is my discipline psychology. Rather, I am a curriculum theorist and my home discipline is education. So it is from my home discipline that I approach this study. Christopher Bollas (1989) remarks that "Each Freudian should also be a potential Kohutian, Kleinian, Winnicottian, Lacanian, and Bionian, as each of these schools only reflect a certain limited analytic perspective" (p. 99). Because my discipline is curriculum theory, the focus of my thinking is not purely psychoanalytic. I am interested in the intersections of psychoanalysis and curriculum theory. The questions I raise pertain directly to lived experience inside schools and universities. I argue that the teaching life and scholarly life are well informed by our colleagues who do analytic work. Of course, analysts understand psychoanalytic work from the inside out. As outsiders to their conversations, curriculum theorists can listen and learn from their writings. Ultimately the aim in this book is to bring back to the classroom, and to my own life in the halls of the academy, insights and understandings that are gleaned from drawing broadly on psychoanalytic literature. The primary aim here is to grapple with what it means to experience the "complicated conversation" (Pinar, 1995) of human subjectivities as they are experienced in the academy and in schools.

CHAPTER 5

Madness, Historicity, and Schooling

Before exploring, in more depth—altered psychic states—I would briefly like to trace historically the concept of "madness" in psychology. It is important to think historically and culturally around the subject of madness and the ways in which psychoanalysis, in particular, has dealt with altered states. I would like to show how historically contingent the notion of psychosis is. An ahistorical and acultural treatment of this subject matter is not only inadequate but is, in a sense, irresponsible. As Philip Wexler (1996) points out, "American academics remain notably unreflexive about the meaning of their work in relation to the larger historical and cultural contexts" (p. 38). The important lesson one learns from Wexler is that when talking about notions of self and or subjectivity, historical and cultural contextualization is necessary, if the study is to make sense in the larger social picture. Otherwise, this study of Otherness might become too narrow and even narcissistic, or just plain uninformed. Decontextualized studies are highly problematic because they tend to perhaps, unwittingly, reify their subject matter. If anything, the concept of "madness" is highly historical and cultural and must be understood in this way, because ultimately "madness" is a social construction and the ways in which one thinks of the concept of madness is always already conditioned by the historical era into which one is thrown. Philip Cushman (1995) comments on the importance of contextualized studies of psychoanalysis. He states,

> In fact, the most common way historians of psychotherapy celebrate rather than critically interpret their subject is by decontextualizing it. By failing to situate the various theories and practices of psychotherapy within the larger history and culture of their respective eras, some historians treat psychotherapy as though it is a transhistorical scene. (p. 5)

A transhistorical approach to any subject matter is suspect primarily because one cannot transcend one's time and place; one writes from a particular sociocultural perspective that must, at the outset, be stated. Curriculum theorists have long criticized the field of education for being an ahistorical field (see, e.g., Huebner, 2000a, b; Pinar, 2000). The drive toward historicization—via the compilation of synoptic texts in curriculum studies—serves as a corrective to this longstanding problematic within the field of education (see Pinar, 2004).

A Brief Historical/Cultural Analysis of Madness

In the foreword to Michel Foucault's (1954/1987) *Mental illness and psychology* Hubert Dreyfus explains Foucault's project:

> [For] Foucault, influenced by later Heidegger, it is no longer possible to speak of mental illness, personality, and psychology as if these notions had an objective reference independent of the practices that give them meaning. What counts as personality and mental illness is itself a function of historical interpretation. (p. xxx)

Although I do not think Foucault's condemnation of psychoanalysis is fair or accurate, Foucault's work—generally speaking—has great value because he demonstrates that seemingly reified terms—like madness—are actually historically contingent. Foucault points out that the notion of "mental illness," for example, is a product of positivistic medical practices. Here, doctors reduce psychic states to organic brain diseases. However, psychotics were not always thought of as "ill," were not always thought of as diseased. Foucault (1954/1987) explains,

> Mental illness has its reality and its value qua illness only within a culture that recognizes it as such. Janet's patient who had visions and who presented stigmata would, in another country, have been a visionary mystic and a worker of miracles. (p. 60)

Recall Ezekiel. He is a good case in point. Considered a prophet by some, but today many might consider him a madman. Historically there is a difference in the way we think about the signifiers "mad" and "ill." Sander Gilman (1994) comments that "the idea of mental illness structures both the perception of disease and its form" (p. 19). The perception of the Other, in other words, is shaped by what we choose to call that particular form of Otherness. The term "mental illness" carries a stigma. To suggest a psychotic is "diseased" has consequences in the way he or she is treated—by other people and by doctors.

Laing (1960) says that "To look and to listen to a patient and to see 'signs' of schizophrenia (as a 'disease') and to look and to listen to him simply as a human being are to see and to hear in radically different ways" (p. 31). R. D. Laing (1960) points out that much "psychiatric jargon" (p. 27) is used to show that the psychotic "cannot measure up" (p. 27). Often, the language of psychiatry is used to show that they (the mad) are not like us (the sane). "They" are considered bad, while "we" are good. Sander Gilman (1994) explains this polarization:

> It is the fear of collapse, the sense of dissolution, which contaminates the Western image of all diseases, including the elusive ones such as schizophrenia. But the fear we have of our own collapse does not remain internalized. Rather, we project this fear onto the world in order to localize it and, indeed, to domesticate it. For once we locate it, the fear of our own dissolution is removed. Then it is not we who totter on the brink of collapse, but rather the Other. And it is an Other who has already shown his or her vulnerability by having collapsed. (p. 1)

It is clear to many commentators that the sane and the mad are not, in fact, two distinct groups. From at least Freud on, many believe that there is a continuum between sanity and madness. Yet this is difficult to think about because it implicates "us"; it turns the tables on our own sense of wellness, health, sanity. As Gilman (1994) points out, it is easier to think that the Other is vulnerable, that the Other is weak, that the Other is mad. Projecting one's own fears onto the other is, in effect, easier, than looking within. Projection, thus, is a convenient psychological mechanism that serves to protect one from interrogating one's own potential madness. It may be un-nerving to think that Melanie Klein (1946/1993) is right!! She argues that everyone has psychotic mechanisms; that the potential to break down, the potential to collapse or psychically disappear is right here, within. It (madness) is not out there (in the Other), It (madness) is in here (in the self). So what if one does fear collapse as Gilman (1994d) suggests? What does this fear signify? According to Winnicott (1992d) the fear probably means that one has *already* collapsed. Winnicott (1992d) suggests that "what we find looks like a fear of madness that will come. It is of value to us if not actually to the patient to know that the fear is not of madness to come but of madness that has already been experienced. It is a fear of the return of madness" (pp. 124–125). The state of collapse resides deeply within. Who does not, especially on a bad day, think about this? And if one has not thought about it, what might that signify? What has already gone wrong?

Prophet, possessed, sick, ill, seer, psycho, schizo. All these signifiers are meant to drive a wedge between the sane and mad and each term connotes denigration, obfuscation, mockery, pathology. Is it irresponsible to say that the mad are not ill, that they are seers and prophets? Must we pathologize madness? Joseph Campbell (1972) suggests, "to my considerable amazement I learned . . . that the imagery of schizophrenic fantasy perfectly matches that of the mythological hero journey" (p. 202). Is the schizophrenic on a hero quest? Watching *A Beautiful Mind*, one might get the idea that Nash was out to save the world—that could be interpreted as a hero quest. There certainly are mythical elements to the phenomenological descriptions one reads in the accounts of, say, a Paul Schreber. Thomas Ogden (1989) points out that the schizophrenic's world is "a world of heroes and villains, of persecutors and victims" (p. 85). If Jungians, like Campbell (1972), suggest that schizophrenic experiences are examples of archetypal phenomena, what does this imply? One of the problems with archetypal interpretations of schizophrenia is that they are ahistorical and do not take into account the variable ways in which phenomena have been expressed over time. An archetypal explanation suffers, I believe, from the assumption that madness is experienced the same throughout history. Archetypal approaches are structural ones whereby similarities are valorized over differences. But as Levinas (cited in Zahavi, 1999) points out, one must be cautious not to reduce differences to sames. Gilman (1994) aptly states that ahistorical explanations trouble. Gilman might suggest that an archetypal or structural approach, "assumes that the perception of illness is constant across space and time, [and] does not reflect the presuppositions of culture, [and] is in no way colored by the associations with the stigma of mental illness (or its glorification), and is basically invariable" (p. 201). At any rate, Campbell's (1972) position is problematic, although interesting. Jacques Derrida (1998b), on the other hand, argues that one simply must call madness an illness, or one is inauthentic and irresponsible.

> It would certainly be disingenuous to close our eyes, either because of some literary feelings or some absentminded politeness, to what Artaud himself describes as a neuropathological persecution. Moreover, that kind of disingenuousness would be insulting. The man is sick. But precisely, how much more naive would it be not to acknowledge this truth: Artaud is telling the truth. (p. 93)

Sickness is a strong word and perhaps the only word that makes sense in the context of Artaud's suffering. Artaud calls himself sick. We must honor what an individual wishes to call himself or herself. Thus, if one says one is sick then we must honor that and not suggest that one is merely on a "hero quest," as Campbell (1972) would have it. Zahavi (1999) stresses that "A phenomenological [or psychoanalytic] analysis of self-awareness obviously has to take the way in which the subject experiences himself seriously" (p. 154). If we are to take Artaud seriously, we must call him what he wants to be called. However, if someone, on the other hand, says she is not sick but psycho, crazy, or mad, we must honor whatever term the individual chooses to use to describe him or herself.

There are, though, other—more nuanced—ways of thinking about signifiers of madness. Sander Gilman (1994) drawing on the work of H. Tristram Engelhardt, says that it is helpful to think of the notion of schizophrenia, for example, as "a pattern of explanation rather than as a disease in itself or as an eidetic type of phenomena" (p. 204). This is an interesting and useful way to think about the historically contingent nature of the term schizophrenia, without pathologizing it. As a pattern of explanation, it attaches itself variously to negative signifiers or not. Kraepelin termed schizophrenia "dementia praecox," Blueler called it "schizophrenia." Louis Saas (1992) explains that today psychiatrists and psychoanalysts conceptualize "Schizophrenia [as] . . . the most severe forms of psychosis" (p. 3). Saas explains that today there are a whole host of terms that attempt to capture the multitude of variation of schizophrenic experiences. Saas (1992) remarks,

> Though particularly common in patients traditionally thought of as schizophrenic, these can also be found, in milder or attenuated form, in other persons—perhaps to a very limited extent, in all of us, but especially in persons having what is called "schizophrenia spectrum" of disorders which include, in addition to schizophrenia proper, the schizoaffective and schizophreniform types of illness and schizotypal and schizoid personality types. (p. 16)

Conversely, some psychoanalysts use none of these terms. Freudians, Lacanians, Kleinians, and Bionians all have their own particular way of speaking about psychotic states. Bion, for example, collapses schizophrenia with psychoses and mostly refers to the "psychotic personality," according to Gerard Bleandonu (2000); Margaret Mahler (1979a) suggests that there are two types of childhood psychosis: autistic and symbiotic. Melanie Klein (1946/1993) refers to schizoid processes as being fixated at the paranoid-schizoid position. What is noteworthy here is that these various signifiers change over time; they are historically constructed.

What disturbs is that this terminology has been used to demonize. Foucault (1954/1987) states that

> our society does not wish to recognize itself in the ill individual whom it rejects or locks up; or as it diagnoses the illness, it excludes the patient. The analysis of our psychologists and sociologists, which turn the patient into a deviant and which seek the origin of the morbid in the abnormal, are, therefore, above all a projection of cultural trends. (p. 63)

The terms "schizophrenia," "psychosis," "madness," are social constructions carrying historical baggage. The question, though, is not really what the phenomena "madness" is called but what the signifier "madness," is used to do. Is this word, "madness," used to romanticize, sanitize, criminalize, demonize, sacralize? Ezekiel the prophet is valorized in the Jewish tradition; his name sacralized. Ezekiel is thought of in a different sense if one calls him "ill" or "possessed." Many of the early prophets were considered possessed because of their wild seizure-like dancing. In fact, Foucault (1954/1987) points out that the term "possession" was popular before the possessed were medicalized and called ill. Moreover, the mad were not always looked upon with fear or hatred. Foucault (1954/1987) teaches that "up to about 1650, Western culture was strangely hospitable to these forms of experience" (p. 67). But no more. There seems to be a connection between the signification of the mad (as tropes of fear and hatred) and treatments of madness. Winnicott is particularly insightful here. Shock therapy and leucotomy are "unconscious reactions to insanity," says Winnicott (1992c, p. 540). Winnicott stresses fear and hatred of the Other is what drives violent treatments such as shock therapy. "Massive guilt feelings and fear and consequent hate are aroused in people who are concerned with mentally ill persons, and I think their unconscious hate also underlay the cruelty to mental patients" (Winnicott, 1992c, p. 540). It is easier to hate what one does not understand than to undo the hate in order to approach Otherness. One cannot approach Otherness if one hates. These cultural fear(s) and hatred(s) against the mad, Foucault (1954/1987) argues, became more manifest, especially during the fifteenth century when the opening of the "great madhouses" (p. 67) occurred. The great madhouses were cruel. The cruelty of the psychiatric tradition is well documented. It is interesting to note that the function of institutions is to institute, literally, a way of Being that is antithetical to Otherness. Both Foucault and Gilman point out the policing function of institutions. Gilman (1994) explains,

> The confinement of the insane in asylums, the real world's counterpart to the ship of fools in art, began in the Renaissance. By the 18th century, the mad were portrayed in the state-run asylum, an image closely associated with other images of confinement such as prison. (p. 24)

Foucault repeatedly suggests, like Gilman, that houses of confinement and asylums were nothing more than penal systems that used extreme forms of torture to "cure." Foucault (1965/1988) comments,

> From the very start, one thing is clear: the Hospital General is not a medical establishment. It is rather a sort of semijudicial structure, an administrative entity which, along with the already constitutional powers, and outside of the courts, decides, judges, and executes. (p. 40)

When madness began to become medicalized, trouble loomed because of the ways in which psychiatrists got entangled with law and had the power to incarcerate. Artaud (1988e) remarks that "medicine was born of evil" (p. 492). Deleuze and Guattari (1987) suggest that "the psychiatrist was born cornered, caught between legal, police, humanitarian demands, accused of not being a true doctor, suspected of mistaking the sane for the mad and the mad for the sane" (p. 120). Winnicott (1992b) contends,

> In these, 1940s, it ought to be axiomatic that mental disorders are essentially independent of brain-tissue, that they are disorders of emotional development. The fact that brain and other physical disease is related to mental disorders does not alter this axiom. (p. 521)

Interestingly, Winnicott ties the psychiatric traditions' bond to empiricism with the cruelty of leucotomy and shock therapy. Sander Gilman (1994) likewise suggests that evaluations of the mad via an empirical lens led to cruel physical treatments. Gilman (1994) argues that "By the end of the 18th century it was commonplace that forms of insanity, such as melancholy, could be identified by the physical appearance of the person afflicted" (p. 26). How does a person "look" crazy? Some of the sanest looking people have done the craziest things. Think of Jeffrey Dahmer. He looked like a nice, clean-cut guy. But he *cut* up and *ate* his victims!! Gilman (1982) comments that "The statement that someone "looks Jewish" or "looks crazy" reflects the visual stereotype which a culture creates for the "other" out of an arbitrary complex of features" (n. p.). Certain proponents of phrenology suggested a smaller skull size proved that one had the "potential for insanity" (p. 33). Gilman teaches that the empirical tradition led to far-flung hypotheses about madness. A case in point is Johann Gaspar Spurzheim. Gilman (1994) explains,

> In 1817 [Johann Gaspar Spurzheim, who coined the term phrenology] who was practicing medicine in London, published a monograph entitled Observations on the Deranged manifestations of the Mind, or Insanity. For Spurzheim (and for later phrenological interpretations of insanity, such as that by Andrew Combe) there existed a clear relationship between shape and size of the skull and the potential for insanity. (p. 33)

Of course today this seems ridiculous, but then again, maybe not. Some not so good enough psychiatrists might make direct correlations between looks and illness. There are plenty of sane people who look crazy and plenty of crazy people who look perfectly sane. The assumption is if you are crazy, you must be treated physically (with torture, in the case of eighteenth century madhouses, or mind altering drugs such as Prozac). Looks = organic brain disease = physical treatment. However, Neville Symington (2002) comments that "particular mental conditions are only diseases by analogy. Too many practitioners are practicing the art of literalism (Doll, 2000). If mental illness is located in the brain, must one get rid of that part of the brain? Must one cut it out, the way one cuts out cancer? Foucault's gruesome examples speak to this point. During what Foucault (1965/1988) terms the classical period of madness,

> At Bethlehem, violent madmen were chained by the ankles to the wall of a long gallery; their only garment was a homespun dress. At another hospital, in Bethnal Green, a woman subject to violent seizures was placed in a pigsty, feet and fists bound; when the crisis had passed she was tied to her bed, covered only by a blanket; when she was allowed to take a few steps, an iron bar was placed between her legs, attached by rings to her ankles. (p. 72)

Physical illness = physical "treatment." Foucault (1954/1987) comments,

> The 18th century had also invented a rotating machine in which the patient was placed so that the free course of his mind, which had become too fixated on some delusional idea, should be set in motion once more and rediscover its natural circuits. The 19th century

> perfected the system by giving it a strictly punitive character: At every delusional manifestation the patient was turned until he fainted . . . A mobile cage was also developed. (p. 72)

Cushman (1995) adds to Foucault's history of horrors. He reports that

> Hermann Boerhaave, a Dutch physician enthusiastically recommended near drowning as an effective treatment. Hidden trap doors in corridors that suddenly dropped inmates into the "bath of surprise," and submersible coffinlike containers with holes drilled in their sides were just two of the many water-torture devices employed in 17th and 18th century asylums. (p. 95)

Louis Breger (2000) tells us that Charcot, "had his patients ingest iron and hung them from the ceiling in iron harnesses" (p. 102). These "medical" practices are so unbelievable as to be laughable. But they are not laughable, they are travesties. One wonders how our medical treatments will be viewed in the future. I have often thought the medical establishment cruel. Many modern day medical procedures seem sadistic to me.

It is well known that reform movements sprang up as early as the eighteenth century because of horrific abuses. Gilman claims that it was the reform movement that was responsible for the medicalization of madness. Medicalization was one way to streamline treatment and make certain that standards of treatment were being followed. Gilman (1994) states,

> The medicalization of psychiatry, by the close of the 19th century was successful. Its success, however, was due to political factors. In Britain, a series of parliamentary commissions began, in the first decade of the century, to examine the abuses of the asylum, abuses that seemed to provide a rationale for its medicalization. "Reform" was simply not enough. For "madness" came to be seen as "mental illness." (p. 183)

Although at the outset, the medicalization of madness was supposed to be humanitarian, it paradoxically worked to worsen abuses to which patients endured. In Bleuler's (1911/1950) classic text on schizophrenia he cites treatments for mental illness that gained popularity in the twentieth century. For example, Bleuler (1911/1950) states that "Lommer and Rohe have again recommended castration" (p. 473). Some preferred "blood transfusions" (p. 476) while others suggested "turpentine injections" (p. 474) for the treatment of psychosis. And of course there is leucotomy whereby the problem is literally cut out of the brain. The horrible truth, as Winnicott (1944/1992) points out, is that some of the doctors who performed leucotomies in the 1940s wanted to "damage the brain" as a way to "cut out the irritating suppressed memories" (p. 527). Louis Sass (1992) explains the medicalization of mental illness this way:

> Much of contemporary "medical-model" psychiatry follows the spirit, if not the letter, of Jasper's approach: treating schizophrenia as a mere epiphenomena of some biological dysfunction or deficiency and downplaying the possibility and importance of seeking a psychological interpretation, or understanding from within, of the experiential world of the schizophrenic individual. (p. 67)

Murray Jackson and Paul Williams (1994) cite Cancro's critiques of biological explanations for psychosis. Accordingly, Cancro states "The biological theories suffer increasingly from reductionism. Psychological phenomena cannot be reduced directly to biological phenomena . . . psychological concepts such as love are not going to be isomorphic with a molecular cluster" (cited in Jackson and Williams, 1994, p. 44).

Why do medical doctors have such a hard time listening to their patients? Doctors feel that patients' narratives are unreliable. Doctors do pay close attention to numbers, to test results, but not to the patient's narrative. Nina Coltart (2000) tells us that doctors who actually listen to narrative accounts (of schizophrenics) are historically new and probably rare. She comments that "It may seem surprising that the revolutionary notion of paying attention to a person who is labeled Mad, and trying to understand him or her, is not much more than 30 years old" (p. 178). Bleandonu (2000) claims that "not until the 1950s was there any real interest in the psychoanalysis of psychotic patients" (p. 106). Before the 1950s Carl Jung and Melanie Klein payed attention to psychotics. In the case of Klein, she even worked out what she called the paranoid-schizoid position. Her important paper "Notes on Some Schizoid Mechanisms" dated 1946 was a groundbreaking piece that gave analysts the idea that maybe they could try to understand what it was that schizophrenics were talking about. David Bell (1999) explains,

> Melanie Klein's work on early infantile experience provided a framework for the understanding of such primitive mental states. Segal, Rosenfeld and Bion analyzed the first schizophrenic patients using ordinary analytic technique. This work has formed the basis for understanding schizophrenic and other psychotic states. (p. 79)

But even before Klein, the paradigmatic change in the way people began to understand madness must be attributed to Freud, even though he did not work with psychotics since he thought that they could not develop transference. Freud's commentary on Paul Schreber's memoirs—a psychotic whom he never actually analyzed—opened up a way of thinking that moved away from organic approaches toward dynamic ones. Zvi Lothane (1992) says that "Of paramount importance is the far-flung heuristic influence of Freud's Schreber analysis and the effect it had on the spread of dynamic approaches among American psychiatrists" (p. 346). Moreover, Lothane stresses that commentary on Schreber has been written mostly by psychoanalysts not of the Jungian persuasion. Why are Jungians not interested in the Schreber document, I wonder? It certainly is one of the most important schizophrenic memoirs in the literature. One could read Schreber's book archetypally and mythically if one were of the Jungian bent. There certainly are many mythical images in Schreber's writings.

Gilman (1994) contends that "Freud and Reik stressed the 'intuitive,' that is, a priori, nature of knowledge, rejecting the sensory, the visual, for a global understanding of the patient represented in their system by the aural" (p. 45). It is the stress on the intuitive, I believe, that makes Freud's writing on Schreber crucial. How can one say what the intuitive is? Andre Green (1999) points out that

> it was because Freud had a taste for reflection, without being imprisoned by the yoke of philosophical concepts, that he was able to take the risk of laying down the foundations of a much madder way of thinking, the intuitive of which came with the invention of the unconscious. (p. 27)

Christopher Bollas (1989) comments that for Bion, too, "psychoanalytic training was an education in intuition" (p. 178). Educational psychologists, unlike those of us doing curriculum scholarship from a psychoanalytic orientation, are perhaps suspicious of the intuitive. Following Ralph Tyler, educational psychologists might feel that what counts is the empirical. Intuition, though, is not testable, verifiable, or measurable. Derrida (1994) calls empirical thinking "vulgar" (p. 61). Curriculum scholars, though, are not interested in measuring data. Curriculum scholarship, like psychoanalysis, is intuitive. When listening to one's inner currents and following the stream of autobiographic patterns one must work, as William Pinar (2004) insists repeatedly, from within. Some of our colleagues in the field of education are troubled by curriculum theorists because they do not understand what we are doing. Some suggest we are too theoretical, too far removed from the schools, too far removed from the everyday. But this is clearly not the case.

Likewise, during Freud's lifetime, many did not understand what he was doing either. The conservative medical establishment had little patience for the "talking cure." Of course Freud was an MD. And this fact complicates the link between psychiatrists (all of whom are MDs) and psychoanalysts (many of whom, at least were initially, lay persons). Freud insisted that lay persons could become psychoanalysts. In fact, he preferred lay analysts to MDs, because he had little respect for the medical community. He simply did not trust them. Further, he wanted to support his daughter Anna Freud who did not graduate from college, yet became one of the most important child analysts of her time. Likewise, Melanie Klein, another important lay analyst, did much to change the face of analysis and, in fact, started a school of thought called object-relations theory. Neither Anna Freud nor Melanie Klein graduated from college. Yet their work became crucially important to the history of psychoanalysis. Interestingly enough, psychoanalysis is one of the few fields that has historically honored women.

Russell Jacoby (1986) reminds us that "the Freudians of the first and second generation were primarily cosmopolitan intellectuals, not narrow medical therapists" (p. 10). But The American Psychoanalytic Association has been historically dominated by MDs until relatively recently. Ernst Federn (1990) reports that "it took twenty years for the American psychoanalytic societies to yield under court pressure and abandon their policy of not admitting non-medical psychotherapists for training as psychoanalysts" (p. xxii). Freud worried that the medical community would ruin psychoanalysis. And it nearly did. At any rate, Deleuze and Guattari (1987) comment on the strange relationship between psychiatry and psychoanalysis:

> In short, psychiatry was not at all constituted in relation to the concept of madness, or even as a modification of that concept, but rather by its split in these two opposite directions. And is it not our own double image, all of ours, that psychiatry thus reveals: seeming mad without being it, then being it without seeming it? (This twofold is also psychoanalysis's point of departure, its way of linking into psychiatry: We seem to be mad but aren't, observe the dream; we are mad but don't seem to be, observe everyday life.) (pp. 120–121)

It is interesting to note that up until the 1960s, according to Sarah Winter (1999), many psychiatrists in America were trained psychoanalytically. Even though this orientation seems to be slipping, still,

Winter, drawing on the writings of Nathan Hale Jr., suggests that "analysts still regularly occupy top posts in psychiatric departments of medical schools" (p. 3). Yet, Freud wanted psychoanalysts to be extricated from medical institutions because he argued that psychoanalysis was an independent, autonomous discipline (much like the way in which William F. Pinar (2004) has argued that curriculum theory is an independent and autonomous discipline).Winter states (1999) that 1902 marks the beginning of Freud's Wednesday Psychological Society. Here followers of Freud would meet and later change the name of this association to the Vienna Psycho-Analytic Society. In 1909 Freud lectured at Clark University in the United States. This is important because it marks the beginning of American interest in the subject. Unfortunately, Jacoby (1986) suggests that "the medicalization proceeded most rapidly in the United States, undermining the cultural and political implications of psychoanalysis" (p. 120). And it was this that Freud feared. He did not want psychoanalysis to fall into the hands of medical doctors. It would be many years before psychoanalysis freed itself from the yoke of the medical profession in the United States.

At any rate, internationalization of the psychoanalytic movement began by 1910 as the "International Psycho-Analytic Association was formally founded" (Winter, 1999, pp. 130–131). By the 1920s the first psychoanalytic journal was founded. Winter says "By the mid 20s, psychoanalysis had an international professional organization and an official journal" (p. 131). But the real turning point in psychoanalytic history is marked by historical events in the 1930s. Jacoby (1986) points out that it was Nazism that changed the face of psychoanalysis. He states "Nazism severed the psychoanalytic continuum. Psychoanalysis eventually prospered in exile, especially in the United States, but it never recaptured its original ethos and scope" (p. 4). The reasons for this vary. Part of the problem, is that American culture differs dramatically from European culture, and the American version of psychoanalysis took on the American flavor. One of the critiques of American versions of psychoanalysis concern the turn to ego psychology. Ego psychology focuses on adaptation and developmental appropriateness. Freud was particularly disturbed by the turn that Alfred Adler (considered the Dean of American ego psychology) took. Jacoby (1997) reports,

> In the discussions in the Vienna Society that preceded the break with Adler in 1911, Freud denounced the revisions. "The whole doctrine has a reactionary and retrograde character." Instead of delving into the

> unconscious, Adler sticks to "surface phenomena, i.e. ego psychology," and succumbs to the ego's own misconceptions. The ego's denial of its own unconscious is transmuted into a theory Freud designated two objectionable features of Adler's work: an antisexual and a reductionist trend. (p. 24)

Mitchell and Black (1995) claim that it was ego psychology that reigned in the American scene until the 1980s. Adam Phillips (1993) points out that it is ego psychology that Michael Eigen critiques. Eigen is influenced by Bion, who in turn was an analysand of Klein's. Phillips (1993) states that "Eigen is a critic of Margaret Mahler's influential developmental theory, and more implicitly of the ego-psychology that inaugurated psychoanalysis in America" (p. xv). Ego psychologists seem to forget the ways in which Otherness is always already instantiated in the unconscious. Rather, they focus on ego strategies to help patients become better adapted to reality.

It is ironic that although Freud was suspicious of Americans and critical of American society, America is the place where most of his work is archived. Harold Blum (1999) reports that there are "eighty thousand items" [of Freud's] (p. 1028) located in the Library of Congress. He says, "the Freud collection at the Library has become the largest collection of Freudiana in the world" (p. 1028). Although many ego psychologists thrive in the United States, there are others, like Eigen and Phillips for example, who honor Otherness, not adaptation. Stephen Reisner (1999) comments,

> This Freud, who adhered always to analysis over synthesis, who preferred puzzles to solutions, and whose unfailing ability to attend to what is left out, contradictory, or uncomfortable in psychic life, left us a psychoanalysis which in significant ways paved the way to the post-modern era. (pp. 1056–1057)

This is an interesting interpretation of Freud to which many do not subscribe (see, e.g., Deleuze and Guattari, 2000). Freud's writings are highly complex. One needs time to dwell in his words in order to understand this complexity. The ego for Freud, takes a back seat to the unconscious, especially in his later work, as he realizes more and more that we are not the masters of our psychic houses. Thus, psychoanalytic work needs to go behind and around the ego to get to the unconscious. The ego only blocks the way. To focus on the ego is to be trapped in the ego's defenses. To focus on the ego is a mistake. Ego psychology took the wrong turn. It is ironic, that the first ego psychologist—one might argue—was Anna Freud.

Despite the return to Freud that Lacan (1955–56/1993a, 1955–56/1993b, 1955–56/1993c) initiated, many worry that psychoanalysis as a practice is a fading dream. The struggles within psychoanalytic institutes could cause the demise of psychoanalysis as a practice. Some even argue that the end has already come for this profession. The number of psychoanalysts across the country continues to dwindle. Douglas Kirsner (2000) states thus: "In the age of managed care and of cognitive behavioral and biological therapies for mental illness, psychoanalysis is generally seen as a 'profession on the ropes' whose hour is up" (p. 1). Marylou Lionells (1999) points out that it is anti-intellectualism that is killing psychoanalysis, the inability to tolerate "the power of the unconscious" (p. 2). I would not doubt this for one second! America has always been an anti-intellectual society. Paradoxically, the university, as I argue in my last chapter, is anti-intellectual at its core. Public education suffers from anti-intellectualism. This is not a new problem, as educationists are well aware. The standardized testing movement is horrifically anti-intellectual. It teaches children not to think. It teaches teachers not to teach. The same issues that plague the psychoanalytic community plague the educational community as well.

Patrick Kavanaugh (1999) worries about the standardization of health care. He states that

> among the consequences of healthcare reformation for the psychoanalytic education has been the increased emphasis upon the development of more standardized programs of study, the predeterminization of educational objectives and outcomes to be obtained, and a predefined and standardized set of core competencies to be mastered by new practitioners. (p. 92)

Curriculum scholars are quite familiar with the problematics of standardization. The rules for health care—the rules that are standardized—are not dissimilar to the dictates of academic accrediting agencies such as NCATE and SACS. NCATE is criminal. Why can't colleges of education not just say no? How can one standardize education or the psyche for that matter? How can one standardize the unconscious? How can one standardize Otherness? Attempts at standardization are not only ludicrous but amoral and irresponsible. Lawrence Jacobson (1999) says that American psychoanalysis has become flattened out and ruined. One might also say that public education has been flattened out and ruined as well. Psychoanalysis (and I would add education) is not "disturbing" enough (Jacobson, 1999, p. 220). Jacobson suggests that in order for psychoanalysis to get back its edge it

> must maintain what I call a robust unconscious: an asocial, nonverbal unconscious, a "radical imaginary." . . . Such a notion of the unconscious gives traction and subversive power to the individual in the social-symbolic arena. (p. 225)

A robust unconscious. What an interesting idea! A robust educational system is what we need as well. How to make our public schools robust. Is the call to the robust merely a pipe dream? Some day we will get back our public schools, we will get back our power as educators. Maybe not in my lifetime—but maybe in the next generations' lifetime—public education will become more robust as curriculum theorists influence more and more teachers and more and more teachers begin to fight for their rights and for their voices to be heard.

As I said earlier, some argue that the psychoanalytic profession is waning. Ernst Federn (1999), for example, argues that this is the case because some believe psychoanalysis has little market value. Conversely, Stephen Mitchell and Margaret Black (1995) argue that psychoanalysis is not doomed, and in fact, "the past decade has witnessed a psychoanalytic expansion of striking proportions" (p. xviii). The expansion, as I see it, is academic. Literary critics and curriculum theorists embrace psychoanalytic theory. Curriculum studies—because much of it is grounded in autobiographical work—is a natural fit for psychoanalytic theory. On the practicing side, though, because a true analysis is so time consuming and expensive, many Americans would rather have short term cognitive or behavioral therapy. Americans would rather take Prozac than discuss dreams.

Resistance to Madness

At this juncture I would like talk about potential psychological resistance to studying the experience of madness. There will be many educationists who probably will have resistance to reading further. Otherness is not a subject for the anti-intellectual. Many educationists are still dreadfully anti-intellectual. More than that though, many educationists might not see the relevance of studying madness in light of the field of education. Many educationists are too narrowly focused on teaching or on the school. I argue that one must study culture to understand teaching and the school. This is not a new point. Dewey argued for this long ago. Studying madness is, in effect, studying culture. Besides there are students sitting in our classrooms who are potentially mad, or who were always already so. Should we not try to understand them in order to be better teachers?

Studying madness throws one into the "unthought known" (Bollas, 1989). Many will not want to go there. However, if "the essence of subjectivity (it's condition of possibility) is alterity" (Zahavi, 1999, p. 112), then scholars might think again about what subjectivity means in the face of the Other. It is crucial that educators think about why they do not want to travel into unknown waters. Deborah Britzman (1998) contends that

> thinking through structures of disavowal within education, or the refusals—whether curricular, social, or pedagogical— . . . engage a traumatic perception that produces the subject of difference as a disruption, as the outside to normalcy. (p. 213)

These disavowals, of course, are many. Some educators want to rely on their own prejudgments of what knowledge is of most worth in the field of education. Some cry "back to the schools." But how can we go "back to the schools," as it were, if we do not know who is sitting in our classrooms! Children suffer from being othered, being teased and punished for being different. How do educators cope with this? Do they even know where to begin? Must we watch yet another Columbine in order to change the way we think about our children? We must get back to the basics—yes. We must study children's sense of Otherness. This is basic to understanding people. If educators can take a moment to understand the most extreme form of Otherness, madness, maybe we can begin to understand Otherness in its less extreme versions—as I have said previously. Eccentricities should not be punished, they should be celebrated. But let us not simply give lip service to the term phrase "celebrate difference." Let us embrace the odd, the eccentric, and the mad by studying what it feels like to be mad, to be odd, to be eccentric without trivializing. Let us begin understanding that which is not rational, that which is unthinkable. The closer one studies subjectivity, the more one realizes that there is no such thing as "normal." We are all a little bit crazy after all! Studying subjectivity, in any deep fashion informs us about the lives of children, teachers, and professors. Subjectivity entails Otherness. And Otherness is strange and highly complex. But one cannot study subjectivity in all of its strangeness in any deep and meaningful sense if one relies on empirical methodology. One cannot do this work thoroughly if one is atheoretical. William Pinar (1998) remarks, "As a graduate student . . . in the late 1960s I complained about the atheoretical character of the

American curriculum field" (p. xv). Not that much has changed in the broader field of education since the 1960s. Yet curriculum studies, since the Reconceptualization that Pinar started in the early 1970s, has been hammering out highly theoretical work around issues related to curriculum. It is through theoretical work that we begin to understand the ways in which lived experience in the classroom is complex and sometimes mad.

A brief survey of the psychoanalytic literature teaches that care and love are not enough when working with patients. I would argue that when teaching education students, care and love are not enough either. In fact, giving reassurance is the worst thing one can do when patients or students are falling to pieces. Adam Phillips (2001) reminds us that Marion Milner

> counsels us to be wary of the preemptive imposition of pattern, of the compulsive sanity of reassuring recognitions of what we might be doing when we are too keen to clear up clutter. Clutter . . . may be a way of describing either the deferral that is a form of writing, or the waiting that is a form of deferral. (p. 71)

The clutter of classroom life should at least be recognized. The classroom is a cluttered place because teachers do not know their students well enough to know what is going on with them. The classroom is psychically cluttered. A good enough classroom is a place where, at least, this is acknowledged. A good enough professor is one who attempts to study what she intuits is going on with youngsters and graduate students for that matter. The most critical issue is that educators begin to take note. Where does a professor or teacher begin to open conversations around what is taboo, or what makes students feel uncomfortable. I suggest educators begin in the discomfort. If anything, professors should not make their students feel at home. The at-home is not where knowledge and understanding grows. Home—is not where we start from, as Winnicott implies. But rather, one might begin where one is not at-home, as Derrida suggests.

A Matrix of Madness

A phenomenology of madness gives readers a better *understanding* of what it is like to be mad, of what it is like to *experience* this extreme form of Otherness. Derrida (1998b) warns that one cannot really "witness" madness, or a history thereof. He argues,

> There is no privileged witness for such a situation—which, moreover, can only ever take the form with the possible disappearance of the witness already at the origin. This is perhaps one of the meanings of any history of madness. . . . Is there any witnessing to madness? Who can witness? Does witnessing mean seeing? Is it to provide a reason [rendre raison]? Does it have an object? Is there any object? (p. 71)

Evocative questions. Troubling questions. How can one witness that which one cannot witness? Even intuiting what madness might be like is not experiencing madness. Can only the mad write their own history because they have suffered it? Or if everybody is mad can everyone write a history of madness? How mad does one have to be to write about madness? Is Paul Schreber (2000) the most legitimate madman in the archives? Can only Schreber write about madness? Is one engaging in voyeurism when examining the mad? Or is it possible to be objective about madness? Does one include the voices of the mad—in one's study—or the doctors' voices only? How to decide? Whatever one does, the archivization of madness is partial, perspectival, made up, socially constructed, and highly selective. Derrida (1998a) is helpful here. He states that

> constant attention to the paradoxes of archivization, to what psychoanalysis (which would not be just the theme or the object of this history but its interpretation) can tell us about these paradoxes of archivization, about its blanks, the efficacy of its details or its nonappearance, its capitalizating reserve—about the radical disruption of the archive, in ashes, without the repression and the putting in reserve or on guard that would operate in repression through a mere topical displacement. (p. 44)

What constitutes the radical disruption of the archive? Perhaps the disruption of the archive is created through its interpretation. Without interpretation, can one approach, say, Artaud or Schreber? Can madness speak on its own terms? Even through the psychoanalytic lens, one cannot fully approach—or understand—either Artaud or Schreber. They remain off limits, finally; they remain at the limits of understanding, at the "boundary situation" (Jaspers, 1932/1971). What does it feel like to be mad? What is it like to be mad? These are phenomenological questions as well as psychoanalytic ones. Can these questions be answered? Probably not. What else remains off limits? Derrida (1998a) talks of the archivization of "off the record" (p. 48).

What is off the record in this study of madness? Is the author's record off limits? Or is the record of the writer's life anybody's open record? Is the public record open to anybody? What is off the record is off the record. But is anything off the record really? Public intellectuals are on the record, more on the record than one would like to admit.

The subject matter of this study is Otherness. The experience of madness takes us to the other side of Otherness. If this study is done in such a way as to honor madness and not trivialize or romanticize it, the off the record is not off the record at all. None of us is off the record. This is so because we all have experienced some form of madness in our lives. One must be sensitive to the "suffering suchness" (Doll, 2000, p. 149) of madness; to the "foundering" (Jaspers, 1932/1971, p. 197) the "stammering" (Deleuze and Guattari, 1987, p. 98). This archive of madness, though, is not the full record and certainly not the entire story. The phenomenological descriptions and excerpts from Artaud and Schreber, for example, are highly selective. Phenomenological descriptions of madness offered by the analysts around which I draw are also highly selective. This study is grappling; it is not meant to be comprehensive in any sense.

Here I would like to explore a matrix of phenomenological/psychoanalytic descriptions of psychotic states of Being. This matrix I have pieced together is meant to unsettle, it is meant to un-nerve. This matrix is meant to disorient. This matrix is meant to be the most extreme trope of alterity; this matrix re-presents the not at-homeness of lived experience. I begin with the question, Whither Oedipus? Psychosis, some might argue, arrives on the scene before the Oedipus complex. Whether King Oedipus arrives early, as Melanie Klein argues, or not so early, as Freud suggests, is not the question. Oedipus simply never arrives; psychosis never meets up with Oedipus. Otto Fenichel (1953b) puts it this way,

> The Oedipus complex has been called by Freud "The nuclear complex" of the neuroses, and we may go further and say that it is in the nuclear complex of the unconscious of mankind in general. Every single analysis provides fresh evidence of this fact, if we accept those cases of extreme malformations of character which resemble lifelong psychosis and in which a true Oedipus complex has never become crystallized, either because the subject's object relations were destroyed root and branch at an earlier period, or because such relations never existed at all. (p. 181)

As Deleuze and Guattari (2000) state, the Oedipus complex is basically mommy-daddy-and me. The child must work through this mommy-daddy affair. But for the psychotic there is no mommy-daddy-and me.

Nobody is there; or maybe there is too much of somebody; maybe there is too much mommy. Margaret Mahler (1979a, 1979b) explains that when nobody is there, no mommy, no daddy and perhaps no "me" or "I" autistic psychosis is at hand. This is a space of Being where the parents and world do not exist for the infant. Here the infant has no sense of subjectivity, no I-ness, no me-ness. The infant might experience—but who knows what. Mahler (1979a, 1979b, 1979c) also suggests that another form of infantile madness is what she terms symbiotic psychosis. Here, in this extreme form of symbiosis, the infant fuses with the mother and has no sense of separation from her. As Beckett said, "It's all symbiosis" (Doll, 2004). Whatever the case may be, Oedipus has not been reached at all. Deleuze and Guattari (2000) state,

> The ego, however, is like daddy-mommy: the schizo has long since ceased to believe in it. He is somewhere else, beyond or behind or below these problems, rather than immersed in them. (p. 23)

But where is the psychotic—psychologically speaking—if not with her parents (or even recognizing that she has parents and that they are separate beings from herself)—or too much fused with her mother so that she has no separate reality or any reality for that matter. Who knows. Where is the infant in that realm of dim Being? Nobody really knows.

Not all psychotics, though, are totally symbiotic or totally autistic. It seems that there is a vast range of experience in the middle that many report, although it seems that symbiosis, in one form or another is a prevalent condition for most psychotics. John Steiner (1999) remarks,

> In some psychotic retreats the rupture with reality may be extreme, but in most retreats a special relationship with reality is established in which reality is neither fully accepted nor completely disavowed. I believe that this constitutes a third type of relation to reality. (p. 88)

This state of thirdness—whereby reality is neither fully accepted nor completely disavowed—may be difficult to understand because most non-psychotics who are not doctors think that psychotics are completely out of touch with reality. And some probably are. But there are other states of madness not as extreme. This thirdness, or inbetweenness that Steiner points to suggests otherwise. The third thing, the third dimension is a neither here nor thereness. Probably, this third dimension is enfolded in an altered state of time and space, as many analysts report. If Heidegger (1927/1962) is right and Being *is* time, what do we mean when we talk about time for the psychotic? From a phenomenological perspective, Zahavi (1999) points out that "a number of phenomenologists stress the intrinsic relation between temporalization and spatialization" (p. 126). If time and space are intrinsically related and time is out of joint for the psychotic, what does this mean? Zahavi (1999) suggests that for people who are normally thought of as having intact egos, time and space feels like this:

> I am the center around which and in relation to which (egocentric) space unfolds itself. Husserl consequently claims that bodily self-awareness is a condition of possibility for the constitution of spatial objects, and that every worldly experience is mediated and made possible by our embodiment. (p. 93)

But what if one is not certain about where one's body begins and ends? What if one has no sense of ego, of I-ness or me-ness? What does this do to one's relation with objects in time and space? What does this do to one's sense of temporality?

What kind of space/time are we talking about in the state of psychosis? Michael Eigen (1993) states eloquently that

> Space and time become bizarre and poisonous playthings, or vacuous. An ecstatic spasm of color may shoot across a dangerous wasteland and for the moment save and uplift the subject's sense of self and other. In another moment, self and other fragment, collapse, spill into, menace, and deplete each other, and possibly vanish altogether. (p. 1)

It is crucial to point out that Eigen and others repeatedly warn that these altered states are scary. Hence, analysts do not, generally speaking, romanticize psychosis, as was done in the Renaissance. Foucault (1954/1987) reminds us that "Shakespeare and Cervantes, at the end of the Renaissance, attest to the great prestige of madness, whose future reign had been announced earlier by Brant and Hiermymus Bosch" (p. 67). There is no greatness to be found in madness. And to suggest this is irresponsible. What analysts find, in most cases of psychosis, is not greatness, but hell. Bion (1994b) reports that "the analyst does not meet a personality, but a hastily organized improvisation of a personality, or perhaps a mood. It is an improvisation of fragments" (p. 74). Bion argues that the psychotic personality is surely "evidence of a disaster" (pp. 74–75). This disaster is perhaps constitutional but also worsened by early environmental failures. Margaret Mahler (1979a) argues that the earlier the damage, the worse the disaster. Preverbal, pre-Oedipal ruin is harder to undo than Oedipal or post-Oedipal damage. In fact, in some cases, there is no undoing of the damage. As Eigen (2001a, 2001b) tells us savage wounds to do not go away. He states, "[M]adness is nursed by trauma when the personality is forming and remains a wound in one's makeup" (p. xx). Schreber (2000) is a case in point.

Sometimes psychotics experience two realities or two worlds simultaneously. Bleuler (1911/1950) and Sass (1992) comment on the notion of "double-bookkeeping." Double-bookeeping is a state of Being whereby one "keeps the books" on two worlds. But what kind of books does one keep? Some psychotics are able to comment on what we might experience as everyday reality (like book keeping), alongside the delusional, hallucinogenic world(s) of which the psychotic is aware. Imagine living in two worlds at once. Imagine being able to keep up with both worlds and report on them! Surprisingly Saas (1992) contends that this state of "double-bookkeeping" suggests that psychotics do know what is going on—at least partially. Saas (1992) states,

> It is remarkable to what extent even the most disturbed schizophrenic may retain, even at the height of their psychotic periods, a quite accurate sense of what would generally be considered to be their objective or actual circumstances. Rather than mistaking the imaginary for the real, they often seem to live in two parallel but separate worlds: consensual reality and the realm of their hallucinations and delusions. (p. 21)

Many analysts suggest that psychotics do know when they are psychotic; that they sometimes are able to distinguish between reality and delusions, that they can tell the difference between a delusion and reality.

Where are psychotics—psychologically speaking—when they disappear into the realm of hallucinations and delusions? Wherever they are, they are deeply inside themselves. Many analysts remark that one of the primary symptoms of psychosis is pathological narcissism (Klein, 1930a/1992; Rosenfeld,1965/2000a, 1965/2000b; Bion, 1967a, 1967b, 1967c; Bell, 1999; Sinason,1999; Symington, 2002; Waska, 2002). Neville Symington (2002) writes, "I have become convinced that narcissism is the core pathology in our contemporary world, the elucidation of which illuminates what we mean by madness" (p. 1). Perhaps in this sense, narcissism suggests the inability to connect to objects or to develop object relations. This is the state of being deeply inside one's self. Narcissism, in this case, is literally a relation to the self, a relation to the interior of Being. But what kind of relation is this? No relation at all, really. A relation implies a two-ness, a relation occurs between subject and object. But here two-ness does not seem to come into play, and if it does, it is hard to tell what kind of two-ness one is talking about especially if this two-ness involves internal objects that are buried somewhere in the psyche. One must ask what it is that keeps the self occupied if not internal objects. Michael Sinason (1999) explains in richer detail the problem of pathological narcissism:

> the "mad self" [as teased out by Martin Jenkins] "had the arrogance and brittle narcissism that is the hallmark of what is known in psychoanalysis as narcissistic character pathology, and which has been described by Rosenfeld (1987) and Bion (1967). This type of character ranges insensibly from the touchy hypersensitivities and egoisms of everyday life at one end of the spectrum, while at the other end it blurs into what Freud called narcissistic neuroses and which are now called psychoses." (Jenkins cited in Sinason, 1999, p. 53)

But perhaps the term "narcissism" does not fully capture the phenomenological experience deeply enough because the term seems to signify merely—at least in the popular imagination—selfishness or solipsism. Of course in the psychoanalytic literature narcissistic pathology suggests something deeper and more complex than just selfishness. But commonsense understandings of the term have misconstrued what analysts have historically suggested by it. What captures narcissistic pathology better is Thomas Ogden's (1989) concept of the "primitive edge of experience" (as the title of his book suggests). The "primitive edge" is what he terms the "autistic-contiguous position." Ogden (1989) explains,

> analytic work with schizoid patients must be informed by an understanding of the way in which schizoid phenomena represent a realm of experience that lies between a world of timeless, strangulated internal object relations and a more primitive, inarticulate, sensory-based world of autistic shapes and objects. (p. 108)

Here self and object merge. Being feels strangulated. This sense of strangulation might be what occupies psychotic personalities. What or whom is strangled? Many accounts of psychosis suggest that when one experiences psychic strangulation, a sort of zombification process takes place whereby the person feels like a living corpse (Eigen, 1996). Artaud (1988a) says, "I am stigmatized by a living death" (p. 92). Michael Eigen (1993) suggests that this "absolute decathexis" is "self-deadening" (p. 105). Strangulated internal objects point toward a too strong and overwhelming death instinct. This is what Andre Green (1999) means by "the disobjectalising withdrawal of investment" (p. 87). Green (1999) explains,

> The success of a disobjectalising withdrawal of investment is manifested by the extinction of projective activity which is translated particularly by the feeling of psychic death (negative hallucination of the ego) which sometimes barely precedes the threat of a loss of external and internal reality. (p. 87)

What would it feel like to not have a sense of external and internal reality? What is a self without borders? Or is this a self with too many borders? Are there too many internal objects strangulating one's sense of an inner and outer world? Too many internal psychic happenings, whatever they may be, prevent one from just Being. What does it feel like to experience a sort of psychic death? I suppose depression, in its most extreme state, feels deadening, suffocating. De-cathexis, or the complete withdrawal from objects, must feel deadening. Without a relation to objects, the world does not matter any more. But the opposite is possible too. If one blends into objects and gets installed inside of objects, one loses one's sense of subjectivity, one's sense of self.

Bleuler (1911/1950) states that the cycle of psychosis worsens when affects begin to disappear altogether. Remarkably, for some psychotics, affects disappear for years on end. Bleuler (1911/1950) contends,

> In the most outspoken forms of schizophrenia, the "emotional deterioration" stands in the forefront of the clinical picture. It has been known since the early years of modern psychiatry that an "acute curable" psychosis became "chronic" when the affects began to disappear. Many schizophrenics in the late stages cease to show any affect for years or even decades at a time. (p. 40)

Imagine not feeling anything for decades!! This is what the death-in-life experience of psychotics must be like: nothing at all. This must be like living in a Beckettian world, where nothing happens twice (see Doll, 2004).To be aware of being in a state of nothing—and yet simultaneously—not be in a state of nothing, pains. Paul Schreber (2000) in his memoir states "Thus arose the almost monstrous demand that I should behave continually as if I myself were a corpse" (p. 135). The attack of the body snatchers, indeed. Richard Alexander (1993) comments that psychotics report that they are "neither dead nor alive" (p. 53). Louis Saas (1992) suggests that psychotics say that they feel like a "corpse with insomnia" (p. 8). Michael Eigen (1996) suggests that analysts need to be cautious when approaching someone who is, what he terms "psychically dead" (p. xxiv). Eigen warns that "For an individual who is used to being dead, a therapist's aliveness may be horrifying" (p. xxiii).

Many argue that psychotics' egos disintegrate, or their egos never developed to begin with, depending upon the severity of the condition. So the question of the ego is a pivotal one when attempting to figure out what psychotics do or do not experience. As Lacan (1955–1956/1993) suggests "The question of the ego is obviously primordial in the psychoses since the ego in its function of relating to the external world breaks down" (p. 144). Bion, though, argues that for many psychotics the ego does not completely disappear. But others suggest that when the ego disintegrates what one meets is a sort of ghostlike figure. Christopher Bollas (1989) says he has met many a ghostlike personality who suffer from psychosis. He states that "the person is really quite gone. The analyst is, then, left with a ghost" (p. 130). Moreover, Bollas states that "the schizoid path taken by the child who develops a relation to these ghosts is an act of alterity" (p. 130). Alterity as a ghostlike state appears in those who are absent. Recall what William Pinar (1975/2000) said about school children who are forced to remove themselves psychically from the classroom. They become absent, ghostlike. Children learn to be absent because this is what the normalization process inherent in schooling demands. It is possible that these children can indeed develop schizoid personalities. States of psychosis arise, of course, for constitutional reasons, but the environment certainly plays a part in worsening an already broken state of being.

When the ego breaks down, so too does symbol formation (Segal, 2000; Waska, 2002). Or maybe, conversely, it is the lack of symbol formation which creates disintegration of the ego. Melanie Klein (1930/1992) is helpful here. She states,

> Thus, not only does symbolism come to be the foundation of all phantasy and sublimation but, more than that, it is the basis of the subject's relation to the outside world and to reality in general. (p. 221)

Symbols allow people to think analogically, to think metaphorically, to think that "this" resembles "that" but does not equal "that," that "this" is not "that." Without the ability to think symbolically, the result is that "this" is "that," I *am* you, I *am* a chair, my analyst *is* my mother, *I am god*. Hanna Segal (2000) suggests that "disturbances in the ego's relation to objects are reflected in symbol formation. In particular, disturbances in differentiation between ego and object lead to disturbances in differentiation between the symbol and the object symbolized" (p. 163). However, some analysts comment that these states pass and that psychotics can sometimes think symbolically (Rosenfeld, 1965/2000a, 1965/2000b; Bion, 1993). But in the throes of "symbol equation" (Segal, 2000, p. 165), "concrete thinking" (p. 165) leads to utter confusion between the self and the other. Concrete thinking is literally thinking literally. If someone says, "you are such an ass," *concretely* this means that you are literally a body part. This rigidity in thought is often remarked upon in the psychoanalytic literature especially in cases of psychosis. One manifestation of this rigidity is that psychotics will often report that they are machines. Michael Eigen (1993) tells us that "a patient, Leila, speaks about her "oblivion machine" (p. 101). Margaret Mahler (1979c) states that "the body image seems thus mechanically put together in a mosaic way, by fragments of a machinelike self image" (p. 191). Louis Saas (1992) comments,

> An understanding of phantom concreteness, and of the reifying, distancing, often self-alienating processes that underlie it, may help to explain why schizophrenic-type persons are so often inclined to experience and characterize their minds in physical or mechanistic terms-comparing themselves, for instance, to machines, computers, cameras, or engines. (p. 95)

Artaud (1998) says he is a "coffee machine" (cited in Derrida and Thevenin, 1998b, p. 38).

One might begin to think that one understands psychosis by a sum of the following characteristics: psychotics suffer from rigidity, the inability to symbolize, the confusion of you and me, mechanistic ways of Being, hallucinations and delusions. But in reality, there is no summing up. Because this is not all there is to it. The oddities just seem to go on and on. For example, simultaneously—alongside rigidity—is also the experience of utter fluidity and falling forever, as Winnicott tells us. Eigen's descriptions of these contradictory states are worth dwelling on, because they defy any attempt at rational understanding. Eigen (1993) contends,

> The psychotic self may approximate moments of absolute fusion and/or isolation. . . . The individual lives in a swamp or vacuum. The self becomes sponge-like and spineless, or brittle and rigid. More often the self goes both ways at once and is confused by its mixture of nettle and putty. (p. 148)

Saas's (1992) comments are equally disturbing:

> Paradoxical though it may seem, these two experiences—self as all, self as nothing—can even coexist at the same moment. During a catatonic period, a patient of mine had the experience of taking off his own head and walking down the vast tunnel of his trachea, where he moved about while examining a new universe composed of his own internal organs. (p. 66)

Yet one reads reports that psychotics feel that they don't have any organs. Schreber (2000) says "I existed frequently without a stomach" (p. 144). Grotstein (2000) comments that Tausk and Federn reported similarly that psychotic patients said that their organs were not there. Or even more disturbingly, Bleuler (1911/1950) points out that the organs are the site of torture! "Any and every organ has been removed, cut-up, torn to pieces, inverted" (Bleuler, 1911/1950, p. 101). Who removes and attacks the organs? Why?

Many psychotics report that they are being attacked by objects or invaded by them. Some are attacked by gangs of objects. John Steiner (1999) explains that "As Rosenfeld (1971a), Meltzer (1961) and others have described, these objects are often assembled into a 'gang' which is held together by cruel and violent means" (p. 8). The reason these gangs are so difficult to undo is because, as Steiner points out, they are "highly structured, close-knit system of object relationships" (p. xi). This close-knit group operates literally like a gang, remarkably like the Mafia. Steiner (1999) remarks that, "the place of safety is provided by the group who offer protection from both persecution and guilt as long as the patient does not threaten the domination of the gang" (p. 8). Schreber (2000) is a good case in point. He talks about being invaded by "240 Benedictine Monks" (p. 57). Schreber was invaded by various creatures who operated as gangs such as "scorpions" (p. 96) and "Jesuits" (p. 97). He was invaded by "satans," "devils," "assistant devils," senior devils," and "basic devils" (p. 26). On and on. Perhaps these are the strangulated object relations about which Ogden (1989) referred. Strangulated objects overwhelm. When one asks where the psychotic is when he seems absent, perhaps he is preoccupied by gangs of internal objects. How does one escape gangs who threaten destruction? How does one destroy what one has created? The psychotic has created this world for whatever reasons, but how to un-create it is the problem. These gangs have a sadistic nature about them, and it is the sadism that is particularly troubling. Much in the literature is mentioned about the connection between a sadistic super-ego (s) and psychosis (Klein, 1929a/1992; Pichon-Rivierre (1947) cited in Rosenfeld, 1965/2000a; Bion, 1967/1993a, 1994b). Melanie Klein (1929a/1992) remarks that "the ascendancy of a terrifying super-ego which has been introjected in the earliest stages of ego-development is a basic factor in psychotic disturbance" (p. 207). This is perhaps what Schreber (2000) means when he repeatedly talks of "soul murder" (p. 33). Schreber realizes that "soul murder" is of his own making. In his own words, he states,

> the crisis that broke upon the realms of God was caused by somebody having committed soul murder; at first Flechsig [one of Schreber's doctors] . . . but of recent times in an attempt to reverse the facts I myself have been "represented" as the one who had committed soul murder. (p. 34)

And Schreber (2000) did not only have one soul, he had hundreds, maybe even thousands of them, thousands of super-egos? Not only this, these various souls "had their own thoughts" (pp. 58–59). Sometimes these souls took the form of "little men" (p. 74). Sometimes the souls appeared as only part-souls. Schreber comments, "quite a number of other instances later I received souls or parts of souls in my mouth, of which I particularly remember distinctly the foul taste" (pp. 86–87). In any event, Schreber says, "a plot was laid against me" (p. 63). Paranoid schizophrenia, no doubt. Super-egos that variously get split off and multiply tend to increase the intensity of states of aggression which are also mixed up with libidinal states as Rosenfeld (1965/2000b) points out, "[P]atients were unable to differentiate between their libidinal and aggressive impulses and their good and bad objects. Both the impulses and objects were felt by the patients to be in a state of confusion" (p. 61).

Let us stand back for a moment and dwell. There is no making sense of these states of mind. Descartes's dream has been demolished!! No clear ideas here. One way of thinking analogically about these experiences is to think about dreams. Dreams are psychotic states. Freud argued that dreams allow the dreamer to work through day residue. But in the case of psychosis nobody dreams the dream (Grotstein, 2000). If nobody is dreaming the dream, nothing can be worked through. Grotstein (2000) declares that, "psychotic illness, in other words, is a testimony to the departure of the dreamer who dreams the dream and the dreamer who understands the dream" (p. 3). Bion (1967/1993a) goes one step further than Grotstein by suggesting that psychotics cannot dream. Bion (1967/1993a) remarks that "the patient now moves, not in a world of dreams, but in a world of objects which are ordinarily the furniture of dreams" (p. 40). Bion explains that the psychotics' *sadistic super-ego* and his subsequent fear of it prevents him from being able to dream. Building on Klein's paranoid-schizoid and depressive-position, Bion argues that because the psychotic cannot work through these positions, he cannot dream. Bion (1967/1993a) remarks that "it is in the dream that the Positions are negotiated" (p. 37). Even if the psychotic has a dream, Bion (2000) remarks, he can do nothing with it. "A psychotic will very often have a dream that has no free associations, so the dream is useless" (p. 37).

Dreams, memories, and thoughts, Bion (1991c) attributes to what he terms alpha elements. If the alpha elements cannot be "digested," (p. 7), if the alpha function cannot process dream elements, then dreams do not work. Bion (1991c) explains that "Alpha elements comprise visual images, auditory patterns, olfactory patterns, and are suitable for employment in dream thoughts" (p. 26). But when the alpha function cannot "digest" (p. 7) these alpha elements, dreaming cannot happen. James Grotstein (1993) points out that Bion's position is counter to what he terms "classical analysis" (p. 7). I find Grotstein's explanation helpful historically because most analysts before Bion had thought that psychosis meant too much Id and not enough ego. But Bion suggests that the Id has been destroyed or damaged so it cannot digest day residue. So for Bion, it is not that there is too much Id, there is not enough of it. Grotstein (1993) teaches,

> Classical analysis has generally assumed that psychosis presupposes an id and instinctual irruptions which are driven by excessive primary process to overwhelm the ego. Bion's notion is seemingly the opposite: a defective and/or deficient alpha function [primary process] which is less able to receive and therefore to dream about the sensory data of emotional experience. (p. 7)

Unlike alpha elements, Beta elements, Bion argues, are like Kant's things-in-themselves. Beta elements equate thoughts with things. Bion (1989a) explains,

> B-elements. This term represents the earliest matrix from which thoughts can be supposed to arise. It partakes of the quality of inanimate object and psychic object without any form of distinction between the two. Thoughts are things, things are thoughts; and they have personality. (p. 22)

Beta elements are in search of a "container," Bion suggests. If they cannot find a container and get "dispersed", they founder, in the paranoid-schizoid position and "become actively depressed-persecuted and greedy" (p. 41). Beta elements are not phenomena, they are more like Kant's noumena but they take on a sadistic character because they cannot find a home in what Bion (1994b) calls "reverie" (p. 53). For Bion, the mother is the seat of reverie. If the child cannot project sensations into the mother, the beta elements have no place to go and become dispersed. The "cohesion" (1989, p. 40) of beta elements is equivalent to Klein's depressive position. If Beta elements are dispersed enough, the ego and superego split and coalesce to form what Bion calls *bizarre objects*. Bion stresses that bizarre objects are not the same as beta elements although beta elements make up the bizarre objects. Bion (1967/1993a) states,

> In the patient's phantasy the expelled particles of ego lead to an independent and uncontrolled existence outside the personality by external objects, where they exercise their functions as if the ordeal to which they have been subjected has served only to increase their number and to provoke their hostility to the psyche that ejected them. In consequence, the patient feels himself to be surrounded by bizarre objects. (p. 39)

The complicated mechanism that produces these psychic moves is what is termed projective identification. Projective identification suggests more than the psychic move of projection whereby the subject projects her hatred onto others. Rather, with projective identification, the subject projects parts of herself along with her hatred *into* the other, not just onto the other (Klein, 1946/1993). Projective identification is the move whereby the self spits out parts of itself and actually lodges these parts *into* another person or *into* an object in order to evacuate the self. Klein (1946/1993) explains that projective identification is the psychological mechanism used to put thoughts *into* things. Projective identification is a defense mechanism used to not think about that which is unthinkable. But this projective identification is not necessarily always a pathological mechanism. In fact, everybody uses projective identification, to one degree or another. However, Klein suggests that it is the "excessive" (p. 9) use of the projective identification that becomes problematic. Klein states (1946/1993),

> In psychotic disorders this identification of an object with the hated parts of the self contributes to the intensity of the hatred directed

> against other people. As far as the ego is concerned the excessive splitting off and expelling into the outer world of parts of itself considerably weaken it. (p. 8)

Klein stresses that it is not only bad parts that get projected, split off and lodged inside objects, one can also project good parts of the self into the other as well. These projections of both "good parts and bad parts" (1946/1993, p. 13) serve to "control" (p. 13) the other and obliterate the distinction between self and other. The purpose of these projections is to ward off the pain of separation and individuation. What is left if the bad and the good are expelled from the self? Does the bad pull the good with it? Who is left? Is anybody left? I wonder. A serious problem occurs when the psychotic cannot reverse the process of projective identification and parts of the self get lost in the object into which the self has been projected. John Steiner (1999) teaches that,

> In many pathological states such reversibility [of projected parts of the self] is obstructed and the patient is unable to regain parts of the self lost through projective identification, and consequently loses touch with aspects of his personality which permanently reside in objects with whom they become identified. (p. 6)

Analysts experience this process of projective identification in what is termed psychotic transference. Here, the analysand feels that she is no longer separate from the analyst. During a psychotic transference, the patient will, as Bion (1967/1993) suggests, "push forcibly" (p. 24) into the analyst. Herbert Rosenfeld (1965/2000a) reports that

> whenever verbal communication was disturbed, through the patient's difficulty in understanding words as symbols, I observed his fantasies of going into me and being inside me had become intensified, and had led to his inability to differentiate between himself and me (projective identification). (p. 77)

A nuanced explanation of projective identification is offered by Betty Joseph (2000). She suggests that projective identification has "three or four different aspects: attacking the analyst's mind; a kind of total invading . . . a more partial invading or taking over parts of the analyst; and finally putting parts of the self, particularly inferior parts, into the analyst" (p. 144). Why would someone do this? Again, many analysts suggest that projective identification is a defense mechanism which serves to evacuate intolerable thoughts. Steiner (1999) terms these moves "psychic retreats" (p. ix); Klein (1946/1993) suggests that psychosis serves to alleviate anxiety. Projective identification produces hallucinatory states. But exactly what these are remains elusive. Bion (1993a) contends that

> hallucinations are not representations: they are things-in-themselves born of intolerance of frustration and desire. Their defects are due not to their failure to represent but their failure to be. (p. 18)

Contrarily, Andre Green (1999) points out that "Freud establishes a close connection between hallucinations and primary process" (p. 166). If this is the case, then hallucinations do function as representations. Green comments,

> Hallucination is a representation, essentially unconscious, which is transformed into perception by being transposed outwards, due to the impossibility of its acquiring an acceptable form for the subject even just within himself. It can only be perceived from the outside. (p. 169)

The function, then, of hallucinatory processes—of which projective identification makes possible—is to deny suffering, obliterate pain along with "psychic reality" (Klein, 1946/1994, p. 7). However, Winnicott (1992a) argues that not all hallucinations are psychotic and not all hallucinations are used to obliterate pain. He states "Most children hallucinate freely, and I would certainly not diagnose abnormality when a mother tells me that in her flat there is a cow in the passage" (pp. 39–40). If someone told me that there was a cow in her flat, I would wonder a little about that! But I get Winnicott's point. Children do make things up, they engage in fantasy, and maybe sometimes hallucinate that a cow is in the flat. Children must be allowed to engage in fantasy or they will not be able to become creative thinkers. Winnicott (1992a) suggests that hallucinations are used to do specific things. Winnicott explains that if there is a cycle of what he terms "dehallucination" alongside hallucination this might signal a problem. Winnicott (1992a) contends that

> Something has been dehallucinated and in a secondary way the patient hallucinates in denial of the dehallucination. It is complex because first of all there was something seen, then something dehallucinated and then a long series of hallucinations, so to speak filling the hole produced by the scotomisation. (p. 41)

Michael Eigen (1996) points out that for Bion, hallucinations do not "always play a negative role" (p. 47). Eigen says for Bion, experiences of "going blank" or hallucinatory states can also lend to a "godlike moment of revelation" (p. 47). Hallucinations, many analysts contend, begin the process of coming back to more integrative states, closer to Klein's depressive position. Rosenfeld (1965/2000b) points out that "Freud (1911, 1924) suggests that many schizophrenic symptoms are attempts at recovery, a concept which I found confirmed again and again" (p. 57).

How does one become, as Eigen (2001) puts it, "anti-hallucinogenic?" I think most of us take for granted that many people are, for the most part, anti-hallucinogenic. Most take it for granted that things make sense, that objects are separate from our bodies, that time moves at a rate one can make sense of, that space seems to make sense, that cows do not pop up in the passage, and that one knows, more or less, what day it is, and whether one is awake or dreaming. What is it about an ego-syntonic state that is more difficult to accomplish than a hallucinatory state? Eigen argues that one must develop a capacity for suffering, "make room for pain" (2001, p. 5) so that self is able to relate to objects. Yet, for the psychotic individual, it is not so easy. It is not just a matter of, as Bleuler (1911/1950) suggests, paying attention to the hallucinations to make them go away. One might think after watching *A Beautiful Mind*, that all Nash had to do was talk to his made up (hallucinated) people who followed him around and his psychoses would just vanish. It is indeed more difficult than this. Eigen (2001a) suggests that the psychotic has no control over her psychosis. Eigen (2001a) states,

> At first a psychotic individual may think he or she can control what is happening, but eventually realizes it is beyond control. Madness can't get rid of sanity, and sanity can't get rid of madness. One sort of consciousness can't totally obliterate another. One is aware of the interplay of "sanity-madness" and develops notions of this interlacing. (p. 46)

Thus, the psychotic feels trapped. Bion (1967/1993a) suggests that the patient "feels imprisoned" (p. 39). Bollas (1989) states that it is at the site of the "ghostline" where ghosts are both "nurtured" and "incarcerate [ed]" (p. 132). Eigen (2001b) might call this "toxic nourishment," where ghosts (Bollas,1989) or strangulated internal objects (Ogden, 1989) get so mixed up that it is difficult to untangle what is the good and bad breast, what is aggression and what is love, what exactly is the toxin and the nourishment, and this is probably why Lacan (1955–1956/1993a, 1993b, 1993c) states that psychotics love their delusions. These ghosts are incarcerated, says Bollas (1989).

If the ghosts are nurtured they will not be so sadistic. When objects which are projected and re-introjected, Klein contends that they get re-introjected in sadistic ways. Klein (1952/1993) states that "The re-introjection of this [bad] object reinforces acutely the fear of internal and external persecutors" (p. 69). It seems that there is no exit. Maybe Sartre had it all wrong. It is not other people who are hell. The self creates its own hell. Once that hell is created, there seems to be no way out.

Schreber, though, suggested that he could reduce his hallucinations by increased intellectual activity such as playing the piano or reading. Like obsessional neurosis, the more one thinks, the more one thinks. The more one thinks, the less one understands one's obsessions. Underneath obsessional neurosis, one may find hysteria, or the witnessing of the primal scene, or a childhood trauma. But these things may be too traumatic to recall consciously. In psychosis, repression is really not at issue, though. What is at issue is using thoughts to *not* hallucinate, to not push the thoughts into the external world. This is what psychotics have to continually work at. Schreber (2000) remarks that "playing the piano and reading books and newspapers is—as far as the state of my head allows—my main defense, which makes even the most drawn-out voices finally perish" (p. 203). Likewise, for Artaud, Paule Thevenin (1989) comments,

> He has to take possession of the whole space, peopling it with signs and words, garnishing it in all possible directions with diverse weapons, without permitting the slightest breach to appear in which malevolent forces could go to work. Not one square centimeter of the notebooks is unoccupied. (p. 24)

Louis Saas (1992) suggests that continual thinking, for psychotics, is an attempt of alleviation of psychosis. Saas terms this problem "hyperreflexivity" (p. 91). He explains that

> hyperreflexive schizoid and schizophrenic patients who attend so intently to their mental processes and other experiences that they transform them, with the result that these processes actually come to be more like states or things. In such a lived-world, conscious phenomena through the very act of being seen, are substantialized into objectlike entities. (p. 91)

Paul Schreber said he suffered from "compulsive thinking" or what he called "continual thinking" (2000, pp. 208–209). Oftentimes, though, these thoughts seemed to come from without. That is, Schreber complained that his thoughts were not his own, but he was being forced to think them. Louis Saas (1992) comments that "patients feel that all their inner experiences are under the control or scrutiny of some other being, or even that someone other than themselves is actually thinking their thoughts or looking out through their very eyes" (p. 22). Bion suggests that psychotic thoughts get projected into objects and then these objects seem to control thoughts. Bion (1967/1993a) explains,

> If the piece of the personality is concerned with sight, the gramophone when played is felt to be watching the patient. If when hearing, then the gramophone is felt to be listening to the patient. The object, angered at being engulfed, swells up, so to speak, and suffuses and controls the piece of personality that engulfs it. (p. 40)

Objects thinking one's thoughts: what a bizarre thought indeed. The mechanism of projective identification is, then, much more than projecting things onto an object.

The flip side of compulsive thinking is what Bleuler (1911/1959) calls "blocking" (p. 32). The function of blocking remains unclear. Yet, Bleuler remarks that compulsive thinking and blocking are inextricably related. These psychotic modes of address form what Bleuler terms "word salad." Schreber (2000) describes blocking as "The system of not-finishing-a-sentence" (p. 198). Not only is thinking interrupted, but it seems that associative links between one word and another become split or lost completely. Lacan (1955/1993c) suggests that the "quilting points" (p. 258) between associative links are destroyed. Bion (1967/1993) writes on what he terms "attacks on linking"; Thoughts do not link up, nor do they make sense. For example, Bleuler (1911/1950) says that "In the hearing of a catatonic, something was said about a fish-market. She began to repeat, "Yes, I am also a shark-fish" " (p. 25). Bion attributes symptoms such as these to early disconnects between the child and her mother as well as the total environment in which the infant is placed. Bion (1967/1993c) contends that

> on some occasions the destructive attacks on the link between patient and environment, or between different aspects of the patient's personality, have their origin in the patient, in others, in the mother, although in the latter instance and in psychotic patients, it can never be in the mother alone. The disturbances commence with life itself. (p. 106)

Sadistic "attacks on linking" become obvious in Bleuler's horrifying descriptions of psychotics' experiences. Bleuler (1911/1950) reports,

> They have ice inside their heads; they have been put in a refrigerator. Boiling oil is felt inside their bodies; their skin is full of stones. Their eyes flicker, as do their brains. They are being plucked as one pulls horsehair out of a mattress. (p. 101)

What is significant here is the sadistic nature of the mind attacking itself; one part of the mind continually attacks another. We see this especially in Schreber's memoirs. Melanie Klein (1946/1993) explains that for Schreber,

> God and Flechsig also represented parts of Schreber's self. The conflict between Schreber and Flechsig . . . found expression in the raid by God on the Flechsig souls. In my view this raid represents the annihilation by one part of the self of the other parts—-which as I contend, is a schizoid mechanism. (p. 23)

Otherness: The Other Side of the Other

Madness, alterity, Otherness. It is not enough to simply toss these signifiers about without understanding them in the context of a rich phenomenology. One can never really understand these states. Psychoanalysts do not understand fully this material. This study's concerns have turned on exploring what it feels like to experience madness. How do these theoretical discussions help to think through or at least begin thinking about the notions of difference and alterity? Do these theoretical explorations look deeply enough into what Bleuler (1911/1950) called "twilight states" (p. 104)?

What do "twilight states" have to do with education? Freud stated time and time again that there is educative value to psychoanalytic study. I do not know how teachers teach well without understanding who is sitting in their classrooms. Maybe our children experience twilight states. Not that all of our children are psychotic, of course. Few analysts would suggest this. But, again, many of our children are not totally without ill-health either. If one thinks of madness as a trope, as a signifier of Otherness, as an example of the most extreme state of Otherness, one may be better able to come back again to other kinds of alterities that are not as extreme and begin to better understand that all alterities are not alike and that all alterities have different historical and cultural re-presentations, and that there is a vast continuum and variation of altered states of being. Dan Zahavi (1999) remarks,

> There are different kinds of alterity, and if one wishes to investigate to what extent self-awareness might be influenced or conditioned by it, it is essential to specify exactly what kind of alterity one is referring to. (p. 195)

I hope the specificity of alterity via madness has been marked in this chapter. One's self-awareness and condition of Being is marked by studying Otherness. I do not understand how teachers can teach anything at all without being grounded in the "unthought known" (Bollas, 1989, p. 18).

A problem that I have been grappling with when studying the notion of Otherness concerns the idea of the absolutely Otherness of the other. The self and the other should, psychologically, be separate and Levinas suggests that the other is absolutely other, for philosophical and ethical reasons. But I have been wondering whether one is absolutely separate from the other, and if one is not completely separate or somewhat separate. How can the other be other? What does Otherness mean if the other and the self somewhat leak or overlap? If the other and the self merge, where is the other? Where is the self? What happens to the notion of difference? The psychotic has collapsed self and other, is no longer other from the other but thinks he is the other person literally. But contrarily, if the self is strangulated by objects, no other exists. And this is what is other about psychosis. If one is composed of internal objects, ghosts, split off parts, breasts both good and bad, it is hard to think that a self is just a self is just a self. This is why the postmodern notion of subjectivity is right on the mark, although I don't find any descriptions in postmodern philosophy as interesting or as useful as in psychoanalysis. For example, Bleuler's (1911/1950) descriptions of Otherness evoke thought. He tells us that

> A depressed hebephrenic saw, in broad daylight, a flock of sheep unaccompanied by a shepherd against an unknown landscape. Three corpses lie there in specific positions, and at the same time the patient's mother is present in the scene to protect him. Blankets are seen lying on a neighbor's roof; a man is being decapitated continuously. (p. 104)

Yes, this is bizarre. This is certainly Other. If one attends to these states, whether dreamt or hallucinated for whatever reason, I bet one would not like what one thinks about. Do we take ourselves seriously? Do we listen to our inner monologues/ dialogues closely enough? Does anyone listen to us? Do we keep ourselves company with our inner voices, as does Beckett (1980)? Do teachers or analysts take a person seriously when she says, for example, I am attending to my inner voices, whatever they may be? And when one is not listening, as in the case of a teacher, what is one doing to the Other, to the student? Shutting others down by not listening is common in teaching. But what if we took students' comments seriously? What if we asked them to take their inner lives seriously? What is it we are asking them to do? Are they prepared to do this? Are our students prepared to listen to themselves deeply? I think Lacan very useful here as he notes that it is important to think seriously about taking thoughts seriously. Lacan remarks (1955–1956/1993a)

> Don't we analysts [I would add teachers] know that the normal subject is essentially someone who is placed in the position of not taking the greater part of his internal discourse seriously? Observe the number of things in normal subjects, including yourselves, that it's truly your fundamental occupation not to take seriously. The principal difference between you and the insane is perhaps nothing other than this. And this is why for many, even without their acknowledging it, the insane embody what we would be led to if we began to take things seriously. (pp. 123–124)

To think seriously—about notions of alterity—takes time; to think seriously about tropes of Otherness takes commitment. To think seriously means to take these subject matters seriously as they are indeed related to our most impossible profession of teaching. Curriculum scholars simply cannot do without seriously consulting their own psychic states. It is not enough to love or to care or to reassure. A good enough professor takes time to consider seriously what Otherness means phenomenologically and psychoanalytically.

PART III

The University

CHAPTER 6

The University as a Trope of Otherness

The Dystopic University

In the first part of this book I examined the trope of Jewish intellectuals as a site of Otherness. In the second part of the book I examined the experience of Otherness by studying the concept of madness. Here, in the final section of the book, I am interested in exploring the university as a site of Otherness. More specifically, I would like to raise questions around how Jewish intellectuals survive chaotic feeling states that are experienced working in what I call the dystopic university. I argue that the dystopic university is a site of strangeness, alterity, and chaos. Scholars in this post–9-11 era are at a loss as to what the purpose of the university is, or where the university is going—philosophically speaking. The dystopic university is a place where scholars do their work—yes—but it is a place of confusion and groundlessness. For Jewish intellectuals, the dystopic university is especially problematic because of the ongoing problem of being Othered—because of anti-Semitism—within academic settings. I argue that it is this very Othering and groundlessness which open spaces for Jewish intellectuals to do their work. The larger problem of the dystopic university, as I see it, is anti-intellectualism. Jewish scholars may grapple with this problem in different ways, of course. But one way to grapple with it, is to become more connected to Jewish issues in one's scholarship.

In this chapter I briefly address anti-intellectualism and then quickly move into a discussion of the notion of the intellectual and what her tasks might be within the site of the dystopic university. In the latter half of this chapter I look briefly at the struggles of Jewish intellectuals who work in universities. I show that anti-Semitism has always been a problem for Jewish scholars. Finally, I outline a phenomenology of the dystopic university. I offer what I consider to be feeling states of chaos that one might feel working inside the academy in the post 9-11 era. One of my arguments in this final chapter is that in order to survive the chaotic place that is the dystopic university one must write inside the chaos and live in dis-order. As Michael Eigen (2001a) suggests, one must "make room" (p. 5) for chaotic states and not repress them.

Curriculum Theorists as Exploratory Intellectuals

Intellectuals who do cutting-edge work, are especially underappreciated or even disliked by conservative academics. Curriculum theorists, much like early psychoanalysts, are suspect. Even though the field of education has been reconceptualized since the 1970s (see *Understanding curriculum*, Pinar *et al.*, 1995), educationists—generally speaking—feel threatened by curriculum theorists. Educationists feel threatened by curriculum theorists because many education professors (who are not curriculum theorists) are insidiously anti-intellectual (for more on the anti-intellectual nature of the field of education see Pinar, 2000b). Curriculum studies scholars work on highly theoretical issues—thinking about what it means to be an educated person in troubled times—despite the dreadful state of public education in this country. Teachers who study curriculum theory and return to their schools might begin to impact the state of public education—eventually. No matter, we must study our field *intellectually* and contribute *intellectually* to our unique academic discipline. William F. Pinar (1994) remarks that curriculum theory is "an intellectually exploratory academic discipline" (p. 52). But the experimental project threatens partly because it continually questions and deconstructs the state of public education. Experimental intellectual work chips away at the horizon, opening out toward the unknown. Nothing is settled. Nothing forecloses on the future. The project is always delayed and deferred because *thinking* Otherwise takes time and is tough. Tough-minded thinking is not soft. Curriculum theorizing, because it is exploratory, is not soft. It takes guts to deconstruct, question, explore, and take risks.

Curriculum studies is a discipline that dreams in the space of *agony* (Smith, 1999) (it is agonizing to do curriculum theorizing in the increasingly corporatized—standard driven—academy), and dreams in the space of *anguish* (Derrida, 1978) (scholars are anguished at the state of the dystopic university and at the state of public schools).Curriculum scholars argue that the work at hand must dare to question and deconstruct the militarization of the school (Saltman and Gabbard, 2003), and Corporate Headquarters University (Readings, 1996). The dystopic university is a site of confusion, chaos, and negativity. It is a place where scholars are pushed underground to do their work, where scholars feel uncertain as to what it is they should be doing. On the other hand, the dystopic university is a place of freedom. Ironically, one might feel freer to express oneself within the chaos of negative spaces. In the face of the symbolic death of the university (for this is what corporatization has brought about), in the face of the death of intellectual culture (the university does not support intellectual work), exploratory ideas must grapple with the face of death. When one faces one's death honestly or the death of one's (fantasy?) culture (an intellectual culture that perhaps never was), one can do one's work honestly and with a sense of urgency. The death of the university, ironically, brings about the life of progressive scholarship. They might try to kill us through corporatization and standardization, but the more they try to kill us the more alive—and perhaps angry—we become, the more we work with urgency, the more we want to get to the work at hand because *everything is at stake.*

Curriculum Theory and Education

Let it be stated for the record, curriculum theorists are not simply dreamers. Since William Pinar's call toward reconceptualization in the early 1970s, the space of the university has changed. William Doll Jr. (personal communication) comments that it would have been unthinkable even ten years ago to call AERA's (The American Educational Research Association) Division B "Curriculum Studies." Around the country, departments of education have changed too. Ten or so years ago, Georgia Southern University instituted an Ed.D. in Curriculum Studies. This is not a traditional curriculum and instruction program where instruction is emphasized, but one where *curriculum* is thought through culturally, historically, philosophically, psychoanalytically, and so on. Moreover, it is not uncommon to see positions for curriculum theorists advertised in the *Chronicle of Higher Education.* During the 1970s, these positions were scarce.

Today the talk is about the (post) reconceptualization. The next generation of scholars are beginning to think about what our field looks like in this new century. The fact that there is a (post) movement attests to the fact that curriculum studies as a field remains healthy and strong. Younger scholars—like myself—are beginning to open out the field in new directions, notably in internationalization, memory text,

Jewish studies, postcolonialism, psychoanalytically oriented studies, chaos theory, cultural studies, American studies, cosmology, and more. Many of these major strands were hammered out by William Pinar in his address at the 2005 Bergamo Conference. The future is here, whether our detractors like it or not. The younger generation of curriculum scholars are taking the field in new directions, whether our elders like it or not. Ready or not, here we come. The 2006 Purdue Conference "Articulating the (Next) Moment in Curriculum Studies: The Post-Reconceptualization Generation(s)" signals the vibrancy of the next generation.

Through study—intellectual work and writing—disciplines do change. Certainly the discipline of education has been radically unhinged, as Derrida might put it, since Pinar's (1994) early essays and the advent of *Understanding curriculum* (Pinar *et al.*, 1995) and recently Pinar's (2001b) magnum Opus, *The gender of racial politics and violence in America: Lynching, prison rape, and the crisis of masculinity*. This is the most stunning intellectual masterpiece in the field today. This book is a testament and a promise to the continued intellectual work in curriculum studies that continues to unhinge the way curriculum scholarship—as it intersects with cultural studies—examines race, gender, and culture. Intellectual work not only changes fields within the academy, it can function to change larger cultural landscapes.

One look at educational titles of books published in the last ten years remarkably demonstrates the explosion of presses which fore-front works on curriculum theorizing. Peter Lang Publishers, in particular, has been extraordinarily instrumental in opening up the field by publishing cutting-edge work in curriculum. Lawrence Erlbaum, Palgrave, Teachers College Press, and others are beginning to open spaces for cutting-edge curriculum scholarship. Indeed, Curriculum studies threatens academic boundaries, foundations, and philosophies by opening up cultural spaces of interrogation. Dreaming curriculum into existence, indeed. But not without the struggle within the dystopic university. There is always already the feeling that the university is troubled by curriculum scholars. The trouble is—we question the very foundation of the university. We question what it is that makes the university tick. We question what our work should be within the university. We seriously interrogate our lived experience within the halls of academe. Curriculum theorists are not the only scholars to write on the university of course, but our writings differ for a variety of reasons which have to do with our own historicity as a field of study.

So let the dream continue. This dreaming will be couched in a series of questions. What is an intellectual? Who are intellectuals? What do intellectuals do? To whom is intellectual work addressed? How do intellectuals do work within the dystopic university? How do Jewish intellectuals, in particular, survive the chaos that is the dystopic university? What does it feel like to live inside the dystopic university?

What is an Intellectual?

Whatever the intellectual is, she troubles. In his Pulitzer Prize winning book *Anti-intellectualism in America*, Richard Hofstadter (1962) contends that "it seems clear that those who have some quarrel with intellect are almost always ambivalent about it: they mix respect and awe with suspicion and resentment" (p. 21). Yes, scholars are resented. The common sense understanding of the work of the scholar is that she does not do anything, that she has a cushy job and does not do real work because she does not work with her hands (for more on this topic see Wendy Kohli, 1999). Yes scholars are viewed with suspicion, especially if they challenge the status quo. Unearth the numbers of professors who were fired for their Marxist leanings during the McCarthy era. Is our era so different in this post–9-11 world? And yet, scholars are also respected by students, who do look up to us as mentors, especially in these troubled times. Resented—respected. Schizophrenic. Dystopic. Which way is the wind blowing? Who knows.

Russell Jacoby (1987) reminds us that the term "intelligentsia" arrived on the scene in the 1860s, while later it became associated with the 1917 Russian Revolution. Revolutionaries = intelligentsia. Revolutions are not started by bread alone; thought is required. The intelligentsia, thus, have always already been a threat. Jacoby (1987) and Robbins (1990) agree that the term "intellectual" appeared in relation to the Dreyfus Affair in the late 1890s. Anti-Dreyfusards were anti-intellectual and anti-Semitic. The anti-Dreyfusards did not shout down with Dreyfus, when he was falsely accused with espionage, but "down with the Jews" (Poliakov, 1985).

In the Jewish tradition, the notion of the intellectual can be traced to the Yiddish term Luftmenshen. Bruce Robbins (1993) explains,

> Like many other writers on the subject, Atlas quotes the Yiddish term "Luftmenshen," or "air man" without visible means of support (25), to describe the principled poverty that is taken to distinguish true intellectuals from the professional journalists, literary scholars, and so on who have come later. Intellectuals seemed to live on air, and thus by poetic

> extension seemed as free as their ethereal movement. As Karl Mannheim had put it in the 1920s, they were "free floating, detached." (p. 7)

The free floating, detached intellectual no longer exists. Perhaps, though, the detached, free floating intellectual was always a mythical creature; perhaps there never really were bohemians after all. How is one detached? One is already embedded in culture, grounded, as it were, in some tradition. No one is really detached and autonomous. A romantic idea, this bohemian.

Bohemians, air men, air women, free floaters, squatters, coffee house junkies, are attached. They are not "air men" (Robbins, 1993, p. 7). They are attached to the culture into which they are always already thrown. Vegans, organics, and other hippie-types are attached and constrained by culture. The West's preoccupation with the autonomous, free individual, that is, the bohemian, who is not restrained in some way—by State, or Church or by corporate culture—baffles. Russell Jacoby (1987) argues that the bohemian is a thing of remembrance past. However, it is interesting that when one visits Bohemian lands (today that would include the Czech Republic), one is struck by a sense of artistic freedom one feels there. In the Czech Republic it feels, at least to a tourist, that one can be who one is and do what one wants. Artists flock still to these Bohemian lands because of the freeness one feels there. Maybe this airy feeling is just a myth, but nevertheless, Prague, for example, feels more free and open than, say, Vienna. Austria is plagued by heaviness, imperialness and parochialism. Vienna, is tainted by its past, it will never again be a home to intellectual creativity as it was during the *Fin-De-Siècle*. Vienna is a site of ruins. How can one be a Jewish intellectual and do any work there? Why would one want to?

The notion of "organic intellectuals," of course, is Gramsci's. Although he realized that some intellectuals work to uphold the status quo, he called for another kind of intellectual who would fight for the rights of the working class. Henry Giroux *et al.* (1995) have continued Gramsci's call by arguing the case for "resisting intellectuals" (p. 653) to fight for the oppressed. Perhaps Giroux's call is too simplistic. To what thing, what creature, do scholars resist? Resist the evil? The evil of what? It is not quite so simple. If scholars are already entrenched within the site of the corporate university, how do they resist what they are already part of? One cannot stand outside capitalist America and pretend to be above it. One is part of it.

And yet, the university is a site of paradox. American institutions of higher learning are sites of aporias. There have always been those within the site of the university who intellectually create, who work to chip away spaces, chip open new horizons of thought in the midst of anti-intellectual zombification. Historian Michael Kammen (1980) points out that Americans are, as the title of his book suggests, "a people of paradox." He suggests that even as far back as de Tocqueville, this paradox was evident, especially to Europeans. De Tocqueville (2000) comments that "there is no country in the world where, proportionately to population, there are so few ignorant and so few learned individuals in America" (p. 55). Like de Tocqueville, Richard Hofstadter (1962) states that, "our history can be considered one of cultural and intellectual conflicts, the public is not simply divided into intellectual and anti-intellectual factions" (p. 19). The difficulty for scholars today turns on the paradoxical nature of institutions. Institutions simultaneously foster intellectual engagement but also encourage anti-intellectualism. I argue that this is the heart of the dystopic university We have to learn how to live within this schizoid place by writing about it and trying to think through it.

Perhaps the notion of the "intellectual" is too static, too male, too much reminiscent of Rodin's thinker. Rather, intellectuals are embodied, raced, gendered, em-placed, sexed. Intellectuals are women too!! Bruce Robbins (1990) points out that "the subject of intellectuals has been about as gender-neutral as pro football" (p. xvii). Whatever the intellectual is, the term signifies "movement" (Pinar, 1994, p. 52) and this movement is embodied. Pinar (1994) points out, "Intellectual identity, like psychosocial identity, is not a frozen phenomenon. There is continuity; there is transformation as well. Such movement is not linear, but diagonal as it were. One moves more deeply into repressed material, integrating what was before dissociated and unconscious" (p. 52). Good intellectual work should not be a site of dissociation. Dystopia is not dissociation per se, although dissociation is certainly part of it. Rather, dystopia means writing through dissociation. Grounded in groundlessness, good scholarly work works to engage. It engages depth and breadth, digging deeper and broader into the gaps and voids, digging into the stacks of one's own unconscious formations; opening up holes that lead one further down and back into the primal space of *ur-ness.* Not to the place of origin, but to the place of Otherness, not to find a whole person, but to discover the fragments, the archives, the arcades within. Searches in the library are also searches within the psyche.

Interestingly, Wald (1987) claims that intellectuals are generalists. Interdisciplinarity is the work of a generalist. Wald explains that "most often an intellectual will be an interdisciplinary generalist as opposed to a narrow specialist or technician" (p. 23). Here he could be describing the work of curriculum theorists who are indeed generalists. Our task is to examine the curriculum writ large, as Pinar (2001b) puts it. We are not subject matter specialists or technicians of teaching. Curriculum scholars analyze historically, theoretically, psychoanalytically, philosophically, and culturally the complex relations between teachers, students, and texts. We research the larger cultural scene into which we are thrown. Our thrown-ness is culturally constrained by tradition. Studying this thrown-ness may loosen up the cultural clamps, chipping away at the horizons of possibility. As Jacques Derrida (2000) emphasizes, scholars must study broadly, culturally, in order to understand the here and now of one's thrown-ness, Derrida says, "A broad analysis (historical, psychoanalytical, politico-economical), and so on, and also somewhere philosophical would be imperative to define this here-and now" (p. 85). Curriculum theorists have been attempting to do just this. John Dewey was a generalist. Dewey and the progressives of the 1920s and 1930s began what William Pinar (1995) terms the "complicated conversation" of curriculum work. There is nothing commonsensical about studying curriculum. To understand curriculum writ large, one must intellectually grapple with culture and tradition, and engage in exploratory studies.

Who Are Intellectuals?

This question is loaded of course. But it is also situated and contextualized within one's own lifework and lifeworld. It is a difficult question to address because everyone's canon(s) of intellectuals differ. So the question might entail a personal response. I certainly do not speak for a generation. The intellectuals upon which I draw are these! Kierkegaard, Camus, Sartre, Merleau-Ponty, and Karl Jaspers. Freud, Melanie Klein, Jean Laplanche, Michael Eigen, Michel Serres, Michel Foucault. Jacques Derrida. New York Jewish intellectuals. William James. Virginia Woolf. William F. Pinar, Mary Aswell Doll, Alan A. Block, David Jardine, Jesse Goodman, Delese Wear, Philip Wexler.

The pragmatists and the poststructuralists, generally speaking, represent entire movements of intellectual work that has impacted the field of curriculum studies for years. Let us for a moment dwell on the connections between pragmatism and poststructuralism because of the import these movements have had on curriculum studies. Richard Bernstein (1997) argues that there are some similarities between pragmatism and poststructuralism. On first glance, readers might not think this is the case. But any careful student of William James will see that he attempted to undermine the idea of foundations and dogmatic standpoints. Poststructural thinkers argue similarly. Of course, the comparison between pragmatism and poststructuralism may not merit justification, since pragmatism was born in America and poststructuralism in France. Differences between these movements abound. Still, pragmatists and poststructuralists agree on what Dewey considered problematics of the "quest for certainty" (cited in Menand, 1997/1917). Louis Menand points out what Oliver Wendel Holmes (another early pragmatist) feared certainty: it leads to war. Pragmatism was born out of World War I; pragmatists feared the endgame of dogmatic ideologies. Similarly, many of the tenets of poststructuralism reflect fears that are generated by today's global terrorism(s).

What do Intellectuals do?

What do intellectuals do? Intellectuals write. Writing, Derrida (1976, 1978, 1981) tells us, is a threat. However, the act of writing—according to the Western philosophical canon—is less important than speech. That is, Derrida explains across his intellectual career that "writing" has taken a back seat—as it were—to "speaking." From Plato through Rousseau, "speech" has been perceived as more authentic and more real than "writing." The spoken word has been historically considered more important than writing because the speaker is present when speaking. And this is the problem of the metaphysics of "presence." The spoken word is what counts because it is present. When I speak to you, I am present to you. You hear my voice.

Conversely, the problem with writing is that the writer is not present when the reader reads the writer's book. One can read what I write without me being there. So writing—because the writer is not present to defend herself—is considered to be more ephemeral and less real because the writer is absent. But let us ask the question again: what is it that intellectuals do? They write. But what to make of books if writing is not as important as speech? Is the philosophical canon anti-intellectual? In a way, yes—if books are seen as secondary to speech.

Derrida explains the problem. Derrida turns to Plato's *Phaedrus*. Here, Derrida (1976) tells us that already in the *Phaedrus*, Plato

> denounced writing as the intrusion of an artificial technique, a forced entry of a totally original sort, an archetypal violence: eruption of the outside within the inside, breaching into the interiority of the soul, the living self-presence of the soul within the true logos, the help that speech lends itself. (p. 34)

The living self-presence of the soul is key here. What is considered real is literally connected to the person who speaks. Utterances are more real than inscriptions, writings. Writing is derivative of speech and once the writer writes the text, that which is written floats off the page, as it were, into a beyond-the-horizon. A written piece can exist without the self-presence of the soul. How can that be? The text takes on its own life, it is a creature that crawls out into the world at its own pace affecting who knows what and by what means. This is what is feared. What truth can inhere in a text without the presence of the speaker? For Plato, Derrida (1981) says, "Writing is essentially bad, external to memory, production not of science but of belief, not of truth but of appearances" (p. 103). For Plato, writers, thus, are sophists, fakes. It is not coincidental that Socrates never wrote anything down. To write is to lie. Socrates, the mouthpiece (speech) for Plato. The voice of Plato? What an obscurantist move. Why did Plato not (speak) for himself? Why insert a character like Socrates into his texts? How do scholars know this is really Socrates? Is Socrates more Plato than Socrates or more Socrates than Socrates? Did Socrates really exist? Where is the historical search for Socrates? Are Plato's writings sophistic? If writing be damned, how did Plato escape his own condemnations? Sophist Plato? Sophist Socrates? Certainly the Socratic method is sophistic/sadistic!! Derrida (1981) claims, "it is above all against sophistics that this diatribe against writing is directed" (p. 106). Derrida (1981) explains,

> The sophist thus sells the signs and insignia of science not memory itself (mneme), only monuments (hypomnemata), inventories, archives, citations, copies, accounts, tales, lists, notes, duplicates, chronicles, genealogies, references. Not memory but memorials. (p. 107)

The more the sophist writes, the more suspect she becomes. More writing means more trickery. Archives are lies, tricks.

In the dystopic university, the more one writes the more suspect one becomes! Universities don't like scholars, they like bureaucrats. Perhaps this anti-intellectual attitude is as old as Plato.

Derrida also points to Aristotle, who, like Plato, felt that writing was secondary, derivative of the spoken word, false, fraudulent. Derrida (1976) teaches,

> If, for Aristotle, for example, "spoken words (ta en te phone) are the symbols of mental experience (pathemata tes psyches) and written words are the symbols of spoken words" (De Interpretatione, 1, 16a 3)

> it is because the voice, producer of the first symbols has a relationship of essential and immediate proximity with the mind. Producer of the first signifier, it is not just a simple signifier among other things. It signifies "mental experience" which themselves reflect or mirror things by natural resemblance. (p. 11)

Philosophy is expressed in writing and yet, Derrida explains, philosophy is paradoxically calling for the end of writing. A good philosophical explanation is supposed to end the debate once and for all. If the debate ends, no more writing is needed on the subject at hand. Proofs for existence of God, if they are good ones, should end the debate on whether God exists. Did St. Anselm end the debate or not? Apparently not. In fact, Anselms' arguments spurred on more philosophical discussion for centuries-to-come. Good writing, like Derrida's, for example, spurs on more writing, not less. But that is not the goal of traditional philosophy, Derrida says. If one gets at the truth of a text, the argument should end. Hence, the paranoid delusional male phantasy—to end debate—if not by words, by wars. Perhaps a symptom of the "crisis of masculinity" (Pinar, 2001a, 2001b). Yet words spur on more words, wars *spore* more wars. Words are always belated, delayed, and almost at once come too soon and too late. Derrida argues that writing and speech are integral to each other. Writing is *not* in fact secondary to speech. Deconstruction, then, is an attempt to overturn the traditional dualism between writing and speech, by showing that both are intertwined and both are important to the work of the intellectual.

Derrida tells us that writing is an "adventurous excess" (1981, p. 54); it is "inaugural, in the fresh sense of the word, that it is dangerous and anguishing. It does not know where it is going" (1978, p. 11). However, Derrida (1981) warns that because writing is uncertain of itself, "The accident or throw of the dice that 'opens' such a text does not contradict the rigorous necessity of its formal assemblage" (p. 54). Yes, writing, good writing, demands a formal assemblage, a footpath, even if it is a zig-zag. These footpath zig-zags are written deliberately, thought out with great deliberation and care.

If we are to look for our historico-mythical archetypes of writing, Derrida (1981) suggests we turn to the Egyptian god, Thoth, for he is the god of writing. "The god of writing is thus at once his father, his son, and himself. He cannot be assigned a fixed spot in the play of differences. Sly, slippery, and masked, an intrigue" (p. 93). Further, Derrida remarks that Thoth "is never present. Nowhere does he appear in person. No being-there can properly be his own" (p. 93).

Thoth is radically thrown all about, running, slipping between this and that. Thoth's marks can be glimpsed in traces of Freud, of James, popping up over time, disappearing, popping up again. Neo-Freudian theory; neo-pragmatism. Speaking again, going silent.

To reiterate a difficult point: unlike Plato, Derrida qualifies what he means when talking of speech and writing by pointing out that these are not separate spheres. Writing cannot exist without speech. In fact, Derrida (1976) contends that one must "recognize writing in speech" (p. 140). In other words, writing is always already encrypted within the spoken word. Speech is not more true than writing, one term does not have more weight than the other. But the tradition of philosophy since Plato, has valued speech over against writing, primarily because writing is a threat. "What threatens is indeed writing" (Derrida, 1976, p. 6). Writing is a threat because it can shift the course of the future, it can shift the course of the university. What is it that intellectuals do? They write. Writing threatens. That is why Nazis burned books.

University as a Trope of Otherness

Institutions of higher education value the voice, not writing. Institutions of higher education seem to over-value the voice of committee work: this is work of the voice. "I am here and declare that." "This meeting is called to order." "Here is the agenda." "Any discussion?" But what is the agenda? What gets discussed? The agenda of service work within the university is that it is opposed to writing, it is contrary to intellectual work, it thwarts intellectual work, it wastes the intellectual's time. It would be better to have meetings via writing. Ah, but that is too threatening. The being-there of committee work has little to do with writing (although writing is always already everywhere) and what does get written down seems, at the end of the day, insignificant in light of the larger calling of doing intellectual work. The work of the voice at committees, always interrupts the real work of the intellectual. Every day, the work of the intellectual is interrupted by the voice at the committee who again calls to meeting yet another meaningless meeting. But the interruptions are, in a sense, not meaningless, because they push intellectuals into doing certain kinds of intellectual work that oppose this Kafkaesque atmosphere. This is where the critique of the dystopic university begins, in the belly of the committee of anti-intellectualism.

William F. Pinar (1994) teaches that the character K. (who is like all of us in the dystopic university) in Kafka's *The trial*—who is nothing but a functionary—responds to his banking position in "self-sabotaging" (Pinar, 1994, pp. 57–58) ways. Burying himself in paper work, burying himself in the trivial day-to-day life, he cannot understand why he is put on trial. Yet he is always already on trial for not living his life, but rather for living the life of the banker, becoming the banker. K. gets taken up in the mundane, only to lose his sense of the larger picture; K. loses his sense of what it means to live a meaningful life. One must never become the institution (a banker is an extension of the bank) or one loses one's soul.

Like K., the scholar who gets caught up in the trivial meaninglessness of paper work, committee work, the scholar who gets fooled by this so-called work, thinking that this is what counts, the scholar who is buried in the "work" that is not meaningful, who hides under the "work," is already dead, is always already the dead scholar. Too many dead scholars and you've got the dead university. The zombie, the machine, the soulless one. Pinar's (1994) insightful work titled "Death in a Tenured Position" speaks to these issues. The scholar's case, like K.'s, is lost. He is guilty of not doing meaningful scholarly work. He is guilty of becoming the banking machine, the robot who no longer knows what thinking is or what the point of thinking is. "The response to one's case is self-sabotaging: increased number of committee meetings, longer hours in the office, more reading, less time writing and thinking" (Pinar, 1994, pp. 57–58). Pinar says that this lull into committee meetings, this lull into functionary existence is indeed dysfunctional and is symptomatic of "psychic deterioration" (pp. 57–58). The dystopic university is the place where psychic deterioration has already happened. Many succumb to deadening life of non-scholarly "work." Doing the work of an intellectual—the kind of scholarly work that takes time and thought and self-reflexivity—threatens the foundation of the dystopic university. And too, good scholarship must work to threaten one's own defense mechanisms. An overly obsessive drive to do committee work, the work of obsessive service does a disservice to the academician because the obsession, which serves to cover over repressed material, works to destroy the possibility of hearing one's writing in one's voice and of voicing one's voice in one's writing. The splitting off of the voice from one's writing is what makes for dysfunctional scholarship, dissociated scholarship, scholarship immersed in projection, and paranoid-schizoid blatherings.

The "teaching first" mantra of many non-research institutions suggests speech over against writing. But how can one be a good teacher if one is not a scholar? What on earth would a teacher teach, if she did not know the current trends of her discipline? "Teaching first" means speech counts, writing is secondary, derivative, and of course suspect. The more a scholar writes, the more she unwrites herself. That is, the more books a scholar produces, the general impression in the dystopic university is that she is not spending enough time working on her teaching. The craziness that is the university. This is madness! In many ways the university is a madhouse.

If a scholar addresses a wider audience in her writing, she is suspect. Presses that open out toward larger audiences, thus, are suspect. Presses that are not university presses are suspect. Vintage is suspect. Is Peter Lang a "real" press? Routledge? Do these presses count in the halls of academe? How about Temple University Press? This press publishes books that, "janitors can understand," according to one of its editors. Who decides what presses count? Is this the public intellectual? Whose press is of most worth?

What is the underneath here? Public intellectuals, like John Dewey, wrote in such a way that language seemed more open, more readable, but not necessarily more understandable. Certainly Dewey wrote in a way that is not jargon-filled, yet I do not find his writing, on the other hand, simplistic or easy to comprehend. Would Dewey be tenured at today's Ivy League Institutions? I wonder. Scholarly communities are meant to write in code. Codes are thought to be technical devices to keep the uninitiated out. If the text appears to be less technical, less coded, less jargon-filled, the judges of scholarly writing find guilty the writing that speaks to a larger public.

The point is that writing is always already suspect. This problem is deeply historical as Derrida points out. In research institutions, writing that is too political; writing that borders on the anarchic is suspect. Russell Jacoby (1987) documents several startling cases of those who were considered to be too radical and were therefore dismissed from Ivy League or Ivy League want-to-be Institutions. Jacoby tells us that Henry Giroux, for example, was not granted tenure at Boston University because he was perceived as too political, too radical. Jacoby (1987) tells us that Paul Starr was dismissed from Harvard the year after he won a Pulitzer Prize in sociology. Imagine that. Win the Pulitzer and get fired! Beware of Vintage Press. Publishing here might make you famous and get you fired too. Foucault, though, made a conscious decision to publish in Vintage. He wanted to be famous. I do think there is a connection between Foucault's fame and the resentment conservative scholars express when talking of Foucault. Those who dismiss his work, probably never read it.

THE DYSTOPIC UNIVERSITY AND ANTI-INTELLECTUALISM

The dystopic university is filled with anti-intellectuals! This is one of the great disappointments of academic life. This fact is paradoxical. Intellectuals and anti-intellectuals are strangely symbiotic. Symbiosis works in strange ways. The more one is surrounded by anti-intellectualism, the more one is driven to become intensely intellectual. A schizoid signification of the anti-intellectual beckons the intellectual to work harder. Similarly, the more anti-Semitic an institution, the more Jewish one becomes in one's thinking. Freud's turn to his own Jewishness was a reaction to working in an anti-Semitic university in Vienna.

In order to talk of intellectual work one must also, schizophrenically, talk about anti-intellectual responses to the thinking project at hand. Intellectual work is not carried out in a vacuum or a sanctuary (Sassower, 2000), but is carried out at the site of a para-site or perhaps at the site of an antithesis, in the house, in-house on-site, in-sight, of an anti-intellectual madhouse. The dystopic university is a madhouse of sorts. Intellectuals are always already living within an anti-intellectual mildew that never seems to go away. It is here in this strange mil-Dewey (the mildew and Dewey are strange schizoid bedfellows) place. This schizoid site creates agony that gloams over the scholar. David Smith's (1999) groundbreaking work on the "pedagon" is particularly instructive here. Smith suggests that intellectual work hovers at the site of an agony, or as Derrida (1978) says—writing—more specifically, is performed in "anguish" (p. 9). Why is this? Because scholars agonize over the crazy conditions within which the work is done. The dystopic university is not friendly to scholars, and certainly not friendly to Jewish scholars. A shocking fact in the twenty-first century, but it is, no doubt a fact. The dystopic university is friendly to anti-intellectualism and to making scholars mad. Of course I mean these things in metaphoric ways, reader. Do not be alarmed at the tropes I use. I use tropes to allow us to think through these difficult times.

At any rate, meaningless interruptions of academic life fill our days. To be an academic means to be interrupted by trivia, adminis-trivia. Hounded by Kafkaesque tasks, academe is ever the Castle (Kafka, 1958) about which Kafka lamented.

Many Jewish scholars do intellectual work as an address of threat. When the Other speaks, threat sounds. These intellectuals threaten those who are anti-intellectual and anti-Semitic. Later in this chapter I will specifically deal with Jewish intellectuals and the site of the dystopic university. But for now, I will comment that, generally speaking, anti-intellectuals within the site of the dystopic university feel threatened when intellectual work ensues. Intellectuals at the site of the university are almost an anomaly! This is craziness, is it not?

Writing is a threat. Who does it threaten? Anti-intellectuals. These come in all stripes and they emerge in all places, most disturbingly in the academy. Historian Michael Kammen (1980) points out that European-Americans inherited anti-intellectualism from their European ancestors. For example, two important European figures who have affected American sensibilities, according to Kammen, are John Locke and William Perkins. John Locke, the father of empiricism, was "characteristically English because of his devotion to common sense, his mistrust of metaphysics, and his doctrine that all knowledge comes from experience" (Kammen, 1980, p. 19). Empirical studies, in my view, are given more weight in the academy, especially in the field of education. Educational psychology is believed to be king of the field. Empirical psychological studies are highly valued. Whose knowledge is of most worth? What knowledge counts? Empirical knowledge. Theoretical work, in the discipline of education, is suspect because what one theorizes about cannot always be literally observed. And what one cannot see (theory) one cannot (hear). No voice present. Derrida is instructive here. The voice, presence, being-there is valued over against writing that treats "mere" thoughts. What one observes is considered more real than what one wonders about. Theoretical scholarship, in educational circles then, is suspect because one does not literally study one's subjects in-the-flesh. One thinks through issues. What scholars in education seem to forget, though, is that in other fields, such as physics, some of the most important work is theoretical (i.e., theoretical physics). Are theoretical physicists suspect? Probably not because they are scientists. The hierarchy in educational studies is clear. Empirical studies are good; theoretical studies are suspect. Empiricism is hard; speculative theory is mushy. As Mary Aswell Doll (2000) puts it, these problems boil down to literalisms. If you can literally observe your subject, you must be on to truth; if you cannot see your subject and observe it—say—if your work is theoretical in nature—it is derivative, secondary, not important, irrelevant.

Michael Kammen (1980) interestingly draws on one William Perkins, an English theologian, to demonstrate how anti-intellectualism can be traced to the Anglican faith. This English anti-intellectual sensibility got injected into the broader Protestant culture of America. Kammen (1980) says, "William Perkins, the great Cambridge theologian whose writings profoundly influenced early New England minds, ranked with Calvin . . . he was a practical rather than a speculative theologian, a Puritan popularizer in search of tangible results" (p. 19). Perkins is not dissimilar from Martin Luther. Luther insisted that in order to be a member of the "priesthood of believers," the faithful needed to strip the Catholic tradition of embellishment, ritual, anything superfluous. What was it that threatened Luther? Embellishment, music, incense, ritual, the philosophical treatment of religion. Writing. For one, he hated the writings of Aristotle because Aristotle was speculative and cherished metaphysics. What embellishes escapes the truth of the pure word. The word is God, is with God. The word is the voice of God. Anything embellishing God's voice is a threat, is suspect. One of the key ingredients to Luther's religiosity is practicality. He wanted to wipe away anything that might lead the believer astray. Certainly, music and elaborate iconography could draw the believer elsewhere. David Smith (1999) recalls that

> Luther, the scrofulous one, the man drinking four gallons of beer a day whose later writings may show signs of paranoid alcoholic psychosis? And what of the downside of the Protestant Reformation as a whole, as it can now be read in retrospect from the point of view of our present experience of it. Its propensity for schism and divisiveness; its ability, in the name of purification, to produce not just intolerance for ambiguity, but also exaggerated imaginal literalism through the death of the symbol. (p. 78)

The death of the symbol—the death of the university—is a call toward practicality, transparency, simplicity. What about these signifiers? What exactly do they mean? What does practical mean? What does it mean to be transparent and simple? Nothing is transparent and simple. So these are the lies that America has inherited from the Protestant tradition. These are the lies that get injected into the American educational system. The death of the symbol, that is the dis-ease that is American public education. There is nothing simple about becoming educated. The call toward practicality is a code for anti-intellectualism. But the call toward practicality is nothing new and it is not only an American problem. Michael Kammen (1980) points out "Practicality, then, is not a peculiarly American virtue" (p. 20). Scholars trace strains of practicality in Locke, Perkins, and Luther, to name but a few.

Kammen (1980) points out that amid the paradoxes of American life, there are two "currents running parallel through American minds: the transcendental current, elevated by Jonathan Edwards, refined by Ralph Waldo Emerson; and the practical current which Benjamin Franklin made into a philosophy of common sense" (p. 111). Interestingly, historian Richard Hofstadter (1962) claims that anti-intellectuals' call toward the practical signals a "widely shared contempt for the past" (p. 238). The search for method in the field of education is ahistorical and anti-intellectual. Courses in "teaching methods" presuppose that students and teachers transcend time and place. How can one have a method for teaching when history changes who we are and where we are in our lives? In the post-Columbine era, we can no longer rely on any clear teaching methodology. Nothing is clear today. Nothing has ever been clear for that matter. We live in dystopic times and the curriculum should mirror this. William F. Pinar (1994) points out that the "traditionalist" who calls for the practical "maintain [s] the atheoretical and ahistorical character of the field" (p. 80). The field of education has always been a hotbed of anti-intellectualism, especially evidenced in Tyler's rationale and Bloom's taxonomy. Intellectual inquiry narrows when educationists think our subject matter turns on the school only. This too is a form of anti-intellectualism. Pinar (2001b) points out this mistake. "While it has been frequently observed that we work in a field called 'education' not 'schooling,' that we are professors of 'education' not 'schooling,' the truth is that the two have often been—and remain—conflated" (p. 18). Dewey's progressive dream entailed the study of culture as it is reflected in the site of education. What is in culture is instantiated already in the university or school. Curriculum studies (which I see as an extension of Dewey's cultural interrogation of education) points toward understanding race, class, and gender as these concepts are contextualized within historically specific sites of oppression. Educationists must turn to the larger cultural, historical, sociological scene to better understand the landscape of curriculum writ large. The parochial "turn to schools" movement is atheoretical and anti-intellectual.

The study of popular culture, or cultural studies, and conferences such as the *Popular culture association of the South* are testaments to the importance of studying issues, educational and otherwise, against a larger—more encompassing—sociocultural backdrop. If we are to understand "youth cultures" as Daspit and Weaver (2000) put it, we must turn to their culture. We cannot understand young people if we do not understand what music they listen to, what books they read, what computer games they play, and so on. Again I return to the issue of Columbine. If teachers and parents might have taken the time to understand these alienated youth, maybe Columbine could have been prevented. When young people feel misunderstood or not heard they turn to violence. The task of educationists might be, in part, to study popular cultural sites in which students find solace. We should listen to our students' expressions of anger, frustration, alienation, and resentment. Let our students teach us; let us not be so arrogant to think we understand them. A student wrote on her midterm exam, "I have learned nothing in school. I learn elsewhere." What does this mean? I have learned nothing in school? If students are not learning in school, they must feel frustrated and angry.

Some education scholars have turned the conversation completely upside-down by examining non-traditional sites of learning. Petra Munro (1989) argues that, historically, educational sites for women have been located outside of universities and schools. Munro's ongoing project of "engendering curriculum history," turns on examining the ways in which women have had to find their own places to get an education. Munro turns to monasteries and medieval mystics, and institutions like Hull House, to better understand women's struggles to become educated—outside of patriarchal culture or on the edges of patriarchal culture. These non-traditional sites of learning served as refuges for women who valued education and learning. It is only in recent history that women have been able to attain Ph.D.'s and land positions within universities.

Young people, too, turn to alternate sites of education because there is little "education" going on in public schools. What goes on in schools is "enforcement," as Saltman and Gabbard (2003) argue. So, to escape the culture of "enforcement" young people turn to comic books, zines, video games, and the Internet. Moreover, youth do body piercing and body alterations—which are symbols of self-expression—against a hyperreactionary culture. Difference in public schools is simply not tolerated, period. Public schools may be the most reactionary sites in America.

Despite the work of curriculum theorists, schoolpeople (who are blind to their own racism, sexism, homophobia, and parochialism) continue to insist that education professors must narrow their discourse to "the school." But how can one narrow one's discourse to "the school" when we need to understand—for example—homophobic responses in a historical way? Do we not need to study homophobia historically and culturally if we are to understand why gay and lesbian students are othered and murdered? Recall the case of Mathew Shepard. It is simply not enough to say we embrace difference. Schoolpeople do not embrace difference, they embrace the "enforcement" (Saltman and Gabbard, 2003) of sameness.

Schoolpeople continue myopic, sexist, racist, homophobic anti-intellectualism that gives education a bad name. Schoolpeople pretend that they stand outside history, culture, and the rest. This is the travesty that is American education.

Shockingly, founders of universities were often anti-intellectual. Richard Hofstadter (1962) points out that both Andrew Carnegie and Leland Stanford were suspicious of the "usefulness" of higher education. Andrew Carnegie "took delight in demonstrating how useless higher education was in business; much as he praised 'liberal education,' he had nothing but contempt for the prevailing liberal education in American colleges" (p. 259). The building that houses the English department on CMU's campus resembles a factory. In fact, it was rumored that if Carnegie-Tech (as it was named before the Mellon's bought part of the university) failed, it would be turned into a factory. It is ironic that Carnegie-Mellon houses one of the most prestigious departments of musical theater in the country. Musical theater is hardly practical! Carnegie would have considered this frivolous, useless, not practical. And yet the musical theater department brings the institution much fame.

Like Andrew Carnegie, Leland Stanford, founder of Stanford University, had little use for a "theoretical education." Hofstadter (1962) says, "Leland Stanford was another educational philanthropist who had no faith in existing education. . . . He hoped that the university he endowed would overcome this by offering 'a practical, not a theoretical education' " (p. 262). Stanford and Carnegie were not alone. Hofstadter comments that in 1881 the Wharton School of the University of Pennsylvania instituted a vocational track in higher education. It is shocking that such a prestigious academic institution, a prestigious research institution would have any truck with vocational education. Schools of business, too, have always puzzled. What are schools of business doing on college campuses? What does business have to do with education? What are Army ROTC's doing on college campuses? Should soldiers not be marching at military schools?

The corporatization and militarization of the academy is nothing new. Russell Jacoby (1987) unearths some interesting critiques of the University of Chicago and the University of Pittsburgh in the early part of the 1900s. Jacoby remarks that "Both Thorstein Veblen's *The higher learning in America* (1918) and Upton Sinclair's *The goose-step* (1923) denounced the heavy hand of business stifling universities. Chapters with titles such as "The University of Standard Oil" (University of Chicago) and "The University of the Steel Trust"

(University of Pittsburgh) composed Sinclair's book" (p. 142). The university, at least since the turn of the century, has been in cahoots with business and the military. Endless debates over the corporatization of the university have been discussed widely in the literature. Donald Macedo (2000) addresses the problem succinctly "Simply put, higher education's raison d'etre is to serve the imperatives of the market and embrace a language that celebrates accountability, privatization, and competition while relegating democracy, ethics and intellectual life to the margins of higher learning" (p. ix). Brosio (1998), Wexler (1996), Pinar (2001b), and Carlson (1998) have similar worries about the corporatization of the university.

Misreadings of Pragmatism as the Call to the Practical

Unlike Taylorism (the factory model of education and a precursor to the corporatization of the University), the Deweyan progressive project was part of a larger vision of the American school of philosophy known as pragmatism. However, Freud (2000/1927) thought that both pragmatism and behaviorism were anti-intellectual. He remarked in a letter to Sandor Ferenczi that "The entire impoverishment of the American mentality has become manifest in pragmatism and behaviorism" (pp. 297–298). Freud had little patience for Americans, even though Americans embraced psychoanalysis. As against Freud, pragmatism, I argue, is not, a call toward the practical, even though William James, C. S. Peirce and John Dewey address practicality in their philosophy. Misreadings of pragmatism abound. Pragmatism needs careful study and close reading. But a surface misreading of pragmatism might lead one to believe that it, too, is a call toward the practical and thus anti-intellectual. Richard Hofstadter (1962) suggests "As a case in point, I have found it desirable to discuss the anti-intellectual implications and the anti-intellectual consequences of some educational theories of John Dewey, but it would be absurd and impertinent to say . . . that Dewey was an anti-intellectual" (p. 22). James, Peirce, and Dewey were hardly anti-intellectual. Their writings are dense, hard to understand, and hardly practical. Pragmatism(s), in their various forms, are actually quite complex. Grappling with the complexity of experience, experience that is not a foreclosure, is the hallmark of pragmatism. But pragmatism, like any other philosophy, is full of contradictions. Thus it is necessary to look at some key contradictory passages to understand why pragmatism is often so misunderstood.

Superficially, one might read the writings of Peirce and James, especially, and find passages implying that the only issue at hand is what is practical. For example, the following two passages taken out of context, I think, are misleading. On the one hand, C. S. Peirce (1997/1904) says, "The method prescribed in the maxim is to trace out in the imagination the conceivable practical consequences—that is, the consequences for deliberate, self-controlled conduct" (p. 56). Like Peirce, James (1997/1907) remarks that "the pragmatic method in such cases is to try to interpret each notion by tracing its respective practical consequences. What difference would it practically make to anyone if this notion rather than another notion were true?" (p. 94). On the other hand, scattered throughout his texts, James suggests that pragmatism should lead to more complexity since nothing can be known in advance. James (1951) remarks,

> The philosopher then, qua philosopher, is no better able to determine the best universe in the concrete emergency than other men. He sees, indeed, somewhat better than most men what the question always is—not a question of good or that good simply taken, but of the two total universes with which these goods respectively belong. He knows that he must vote always for the richer universe, for the good which seems most organizable, most fit to enter into complex combinations, most apt to be a member of a more inclusive whole. But which particular universe this is he cannot know for certain in advance. (p. 180)

James's interest in contingency and indecision are scattered through his texts. He was alarmed by the dogmatism of behaviorism and other forms of thinking that closed off possibilities. James suggested that one must choose carefully and deliberately between what he termed "live" and "dead" options before making decisions. James's work on memory and consciousness are quite complex. What is practical about memory? Memory is quite elusive. Consciousness is a highly complicated subject as well. There is nothing practical about thinking about what it is that makes us think. What is practical about memory or consciousness? If anything, these discussions are not "useful" (practical); they are musings, poetics on states of mind. Like James, Dewey(1997/1917) remarks that the point of doing philosophy under the sign of pragmatism is to think in a more complex manner. "[T]he pragmatic theory of intelligence means that the function of the mind is to project new and more complex ends" (p. 228). Practicality, on the other hand, reduces complexities to simples.

Like James, Peirces's writings indicate that thought was born, not of simplicity and dogmatism, but rather in what he called "hesitancy."

There is nothing practical about hesitating, in fact "hesitating" is quite complex. Peirce (1997/1878) comments that intellectual movement comes via "doubt" and "indecision." "Feigned hesitancy, whether feigned for mere amusement or with a lofty purpose, plays a great part in the products of scientific inquiry. However doubt may originate, it stimulates the mind" (p. 31). Of the pragmatists, I think that Peirce is the most obscure. What is practical about obscurantism? Like Peirce and James, Freud's writings are filled with contradictory statements and continually changing positions (Morris, 2001). He is never clear or easy. Neither are the pragmatists. Because thinkers change their minds and argue from contradictory points of view, however, does not merit dismissal out of hand. For me, contradictions point to a thinker's struggle to sort through the chaos and order of experience. There is chaos and there is order and that is the way the world is. Ideas should reflect this. However, the very name pragmatism sends, I think, a misleading signal. Dewey points out, in response to much criticism of the movement, that the practical and the pragmatic are *not* the same. Dewey (1997/1917) contends,

> Many critics have jumped at the obvious collusion of pragmatism with the practical. They have assumed that the intent is to limit all knowledge, philosophic included, to proximity "action," understanding by action either just any bodily movement. . . . James's statement that general conceptions must "cash in" has been taken (especially by European critics) to mean that the end and measure of intelligence lies in the narrow and course utility which is produced. (p. 227)

Serious students of pragmatism understand that it is about more than decision making and practicality. Every thought and action is based on a contingent set of beliefs. Beliefs are not foreclosed or predetermined in any way. This was the very problem with behaviorism. Behaviorism had it all figured out. The world seemed clear and science seemed easy. Its purpose was to predict and control. In a world that seems orderly and not chaotic, behaviorism seems to work. But the world is not merely orderly. It is radically chaotic. Pragmatism, in general, was a reaction against determinism and problem solving that reduced the complex to seemingly transparent simples. Summarizing C. S. Peirces' work, Richard Bernstein (1997) comments that Peirce "sought to demolish the idea that there are or can be any absolute beginnings or endings in philosophy. He sought to exorcise what Dewey later called "the quest for certainty" and the "speculative theory of knowledge" " (p. 386). Bernstein also points out that Peirce's theory of signs did not foreclose on the future; hermeneutically speaking, his theory opened out toward an excess of interpretation. Perhaps this is why there seems to be growing interest in Peirce today. Thus, the Deweyan progressive project which curriculum theorists are trying, as Pinar (2001) says, to "resuscitate," is part of a larger American movement of thought steeped in paradox and complexity. If anything pragmatism helps one think through issues that open out to more thought, more interpretation. I hope I have made it clear here that practicality and pragmatism are two different things, as Dewey pointed out. But one can see how these two terms can easily be conflated. I am a big fan of pragmatism and cringe when my intellectual mentors such as Dewey or James are maligned for being simpleminded, which they are not.

Students of Pragmatists: Jewish Intellectuals housed in Universities

Some of the pragmatists, especially William James, influenced many Jewish intellectuals at Harvard. It is to these Jewish intellectuals I turn. Here, I am trying to understand what role Jewish intellectuals have played historically within the halls of academe. One must realize that Ivy League Institutions have been hostile to Jews (Jacoby, 1987; Ritterband and Wechsler, 1994; Morris, 2001). But some Jews did manage to get into Harvard and study with William James and other pragmatists. Milton Konvitz (1994a, 1994b) comments that Morris Raphael Cohen and Horace Kallen are two names that do not often come to mind when thinking about the who's who of New York Jewish Intellectuals who attended Harvard or other Ivy League institutions. They are, As Carole Kessner (1994) puts it, the "other" New York Jewish intellectuals, the intellectuals who usually get left out of the canon. Both of these distinguished scholars left Harvard as students of William James, as students well prepared for positions in higher education, but anti-Semitism prevented them from landing posts immediately after graduate school. In the case of Morris Raphael Cohen, Konvitz (1994a, 1994b) tells us that he "left Harvard with letters of recommendation from William James, Royce, George Herbert Palmer, and Ralph Barton Peny. He also had such letters from Felix Adler and William T. Harris . . . Yet he found no open door, no welcome" (p. 132a). It took Konvitz six years to land a position.

Another student of both James and Santayana was Horace Kallen. Konvitz (1994a) explains that "After graduating from Harvard . . . he remained an instructor in English at Princeton, where he remained for two years. When his contract was not renewed, it was intimated that had it been known he was a Jew, he would not have been appointed in the first place" (p. 146). These facts should not shock, but they do. Official or unofficial anti-Semitic policy reigned supreme at most Ivy League Institutions for the better half of the twentieth century. Traces remain and the trend continues, perhaps in more insidious, unofficial ways.

Those Jewish intellectuals who did land posts at institutions of higher education tended to begin as radicals but toward the end of careers, many—surprisingly—became conservatives. Sidney Hook comes to mind here. Alan Wald (1987) points out that Hook, in 1933, as "assistant professor of philosophy at New York University, wrote the political program for a new revolutionary communist party" (p. 3). In 1980, Hook "campaigned for Richard Nixon" (Wald, p. 7). Wald (1987) and Jacoby (1987) suggest that many New York Jewish Intellectuals became conservative at some point in their writerly lives. Wald contends that the turning point came during the Moscow Trials. Many of the New York Jewish Intellectuals were Marxists. But the Moscow Trials made many re-think their Marxist leanings. Alan Wald (1987) points out that these trials resulted in "a mass purge . . . millions of workers, peasants, party members and government officials were arrested and executed" (p. 128).

Jacoby (1987) argues that this was not the only reason why many Jewish intellectuals became conservative in their later years. He claims that "No dense Freudian theory is necessary to explain that economic deprivation and cultural estrangement often led to an identification and over identification with the dominant culture" (p. 90). Jews have always lived in a constant state of alienation in America. When one feels alienated, othered, different, the move toward assimilation seems to ease alienated discomfort. But traces and remnants of alienation get introjected within the psyche and only return later as a symptom of failed repression. Assimilation is a symptom. Jewish life is always one of alienation. With the recent events of September 11, one worries about the always already anti-Semitic hostility that ebbs and flows across every generation of American life. Those already hostile to Jews, might search for a scapegoat in this latest tragedy. Jews have always been blamed for everything; we have been blamed for the plagues of the Middle Ages and for the stock market crash of 1929; from the rise of Bolshevism to the fall of the Soviet Union. It is no wonder that many Jews make radical shifts, schizophrenic shifts from left to right. How does one manage scapegoatism? The loss of radical Jewish thinkers though, is a problem about which the left needs to take seriously. Where is our current Frankfurt School?

Arthur Goren (1994) tells us that Ben Halpern, another Jewish intellectual not often associated with the who's who of Jewish intellectuals, comments that "there was [and is] an ideological and historical barrier that prevented and prevents full acceptance of the Jews. Culturally, America was [and is] a Christian country, and neither Christian Americans nor American Jews, no matter how tenuous their religious ties, could cast aside the theological folk legends that defined the separateness of the Jew" (p. 72). Thus, Kessner (1994) emphasizes that "It is this 'sense of apartness' [as historian Martin Malia, says Kessner] that is theme to the variations of almost every attempt to describe and define the Jewish intellectual, beginning with Thorstein Veblen's emphasis on marginality" (p. 5). American and European Jews share this sense of alienation, as I pointed out in earlier chapters of this book. In a stunning, eye-opening book titled *Kafka: The Jewish patient* Sander Gilman (1995a) argues that Kafka's work is better understood if one examines the larger cultural conditions out of which Kafka wrote. He was a Jew and he was alienated. Gilman's assessment is correct: Kafka is the Jewish patient. The bug in *The Metamorphosis* (1915/1972) symbolizes Kafka's feelings of alienation as a Jew in a hostile anti-Semitic culture. Psychological and even physical shape-shifting is necessary when one attempts to battle hatreds. Jews attempting to assimilate drop traditional garb, change names and avoid speaking Yiddish in public. These alterations in self-presentation have got to affect the psyche. How can one's psyche not be split off when one constantly manipulates look, speech, and dress?

Broadly speaking, the New York Jewish Intellectuals were not associated with academic institutions before 1940 (Jacoby, 1987). In fact, Jewish intellectual life, according to Harold Stern, says Jacoby, "excluded professors" (pp. 16–17). But after 1940, it became increasingly difficult to be an intellectual without being housed in an academic institution. Even the *Partisan Review*, the engine of the New York Jewish intellectual movement, was eventually "passed into university hands, its editors largely English professors" (Jacoby, 1987, p. 74). The difference between the New York Jewish Intellectuals and the pragmatists (and other kinds of intellectuals who lived and worked during the early part of the twentieth century) is that the New York Jewish intellectuals formed a cohesive group and lived in New York City. Carole Kessner (1994) sums this up.

> Remarking that although American intellectuals, including the Transcendentalists, had done their work mostly in isolation one apparent exception is the group of writers . . . who mostly had been resident

> in New York in the 1930s and who rose to prominence in the mainstream American intellectual life in the 1950s. The group primarily cohered around Partisan Review, which held the view that it was not only possible, but also natural to unite aesthetic avant-gardism with political radicalism. (p. 3)

The list of New York intellectuals is well known: Saul Bellow, William Barrett, Alfred Kazin, Lionel Trilling, Sidney Hook, and Irving Howe, to name but a few. Jewish intellectuals who are housed in academic institutions cannot not feel disoriented. Jewish intellectuals who are housed in academic institutions understand the notion of dystopia because it is a dystopic condition to live inside of a place that has historically been anti-Semitic. Jewish intellectuals feel much the way psychotics feel. They live in a double-universe, do double-bookkeeping (Bleuler, 1911/1950), feel paranoid (Klein, 1946/1993), long to escape into fantasy worlds (Gilman, 1995b), search for reparative pedagogies (Britzman, 2003; Sedgwick, 2003). There is much talk in the Jewish community about repairing broken shards (Wexler, 1996). *Tikkun* means mending the world. Jews want to repair a broken world. Jews are always already trying to mend broken psyches. But some damage is far too serious to mend. In a book titled *The sensitive self*, Michael Eigen (2004) comments that

> disaster does not go away. Heart attacks, cancer, terrorist attacks, emotional and physical and economic abuse—feeds disaster anxiety. Our psyche partly forms around an internal sense of disaster that links with rich arrays of disaster fantasies. (p. 9)

Jews live disaster. Living disaster means living crazy. We live disaster because it is a disaster to be Other in American culture. This is *not* a tolerant culture. The university is not tolerant of difference. Not only this, Jews live disaster not only because of being Othered, but they live the disaster that is university life in general. Universities are dystopic places where things are so chaotic and unstable that one cannot get one's bearing on anything. Jews have a double-disaster to contend with. It is hard enough being a professor in a university and surviving an anti-intellectual climate. But it is really hard being a Jew inside of academe. How can one not be paranoid? Eve Sedgwick (2003) calls for a move beyond paranoia, drawing on the work of Melanie Klein. But realistically how can one *not* be paranoid? Are there not real reasons why Jews becomes paranoid? As against Sedgwick, I would wonder what is going on with a Jew who is *not* paranoid. One way to read Melanie Klein (1930/1992; 1946/1993) is to suggest that one is always already simultaneously paranoid while searching for the reparative. The two positions are fluid and in motion throughout one's life. A more conservative reading of Klein might suggest, like Sedgwick's, that one must move beyond the paranoid to a more reparative phase of life.

Intellectuals against Academe: Schizophrenic Desire

During the 1930s and 1940s, intellectuals could and did exist outside of academic institutions. Intellectuals, for example, could be found at publishing houses in New York. One trend did surface among some of these New York intellectuals working at publishing houses: contempt for academe. It was thought by some intellectuals that the academy ruined creativity, that it was hostile to radical thought, that it worked only to preserve the tradition.

Mary Louise Aswell, well-known fiction editor, is a case in point. Mary Louise Aswell, fiction editor at *Harpar's Bizarre* from the 1930s to the 1940s discovered and nurtured the careers of many great writers from Truman Capote to Eudora Welty to Carson McCullers. In her diary dated 1932, she mentions James Agee arriving at her apartment "drunk for twenty four hours." Friends with Alfred Kazin's sister, Pearl, and William James's nephew, Bill, Aswell represented a generation of editors who lived the intellectual life in New York City during the 1930s. Play readings, soirees, intellectual discussions of "prehension," (could it be Whitehead?) intrigued Mary Louise and others in her group. Writers and editors in New York had their own clubs, their own intellectual groups.

Mary Aswell Doll, Mary Louise Aswell's daughter, recalls her mother scornfully attacking the academy. That is not the place where intellectual work gets done! And yet, Mary Louise took courses at Yale and Harvard where she studied literary criticism with Professor I. A. Richards. Duncan Aswell, the son of Mary Louise, remarked in her obituary that while in Europe, "She met Gertrude Stein, and saw James Joyce and Ezra Pound" (1985, p. 6). One cannot live the life of an intellectual in those stuffy halls of academe! No, one should travel abroad to hobnob with great writers, the true intellectuals of our time.

Like Aswell, Walter Benjamin found that intellectual life teemed on the streets and arcades of nineteenth century Paris. Jacoby (1987) says that "Walter Benjamin mused at the relationship of 19th century Paris—its streets and arcades—to the new intellectual types, such as the man of letters who wandered about, retiring in the afternoon to a cafe to write cultural fillers for the press" (p. 30). Benjamin was a true man of letters, a true intellectual who lived like a Bohemian. Like Benjamin, Jack Kerouac represents American Bohemianism. This bohemian fiction writer lived on the road, working at various jobs along the way. *On the road* (1957) is the classic American bohemian novel. Drunk with life and travel, Kerouac and his hipster buddies travel around America writing, drinking, and living hard. But are there bohemians anymore? Is the idea Bohemian simply romantic? Today intellectuals cannot be bohemian unless they are independently wealthy. Most intellectuals today are attached to universities. Living inside a university is a schizoid experience. We do not want to be there (because of all the craziness of anti-intellectualism), yet we have to be there.

Schizophrenic desires of intellectuals should be taken seriously. The academy's attraction and repulsion is part of intellectual life in the twentieth and twenty-first centuries. Perhaps the streets of Paris or Route 66 remain the site of true intellectual exploration. But who can afford to live on the streets of Paris!! And who wants to live on the interstate? One must have support to do one's work. Let us be real here. Unless one is extraordinarily well off, finding a place to do intellectual work means attaching oneself to an institution. One must be institutionalized. Ah, here's the rub. William F. Pinar (2000b) points out, "Intellectual life cannot now, at least in the United States, be easily or sharply separated from institutional life" (p. 43).

Doing Intellectual Work in the Dystopic University

English and German universities during the eighteenth and nineteenth centuries were tied to both the Church and the State. Today, universities are tied to business and as Bill Readings (1996) puts it "the techno-bureaucratic notions of excellence" (p. 14). Michael Hofstetter (2001) tells us that the Oxbridge model, according to Sheldon Rothblatt, introduced "a new Idea of a university . . . that universities even had an idea at all" (p. x). English universities "were first and foremost seminaries for the Anglican Church" (Hofstetter, 2001, p. 3). Theology and philosophy were the crowning disciplines of the university. Toward the turn of the century English universities began to rank the study of literature at the top of the academic hierarchy. Bill Readings (1996) explains: "In the late nineteenth and early twentieth century, the English, notably Newman and Arnold, carried forward the work of Humboldt and Schlegel by placing literature instead of philosophy as the central discipline of the university" (pp. 15–16). Although English is considered to be a prestigious discipline today, there are few jobs listed in the *Chronicle* for English professors. Today, philosophy and theology are at the bottom of the academic hierarchy. Many philosophy professors fear that departments of philosophy will disappear eventually because philosophy is not "useful," not tied to the market. There is no purpose for philosophy any more. Who needs philosophy when we've got Enron and Halliburton? Who needs philosophy when colleges of business bring in tons of money? Shades of Carnegie and Stanford. Philosophers of education have been disappearing into foundations and policy departments. Philosophy of education is a dying field. Educational foundations' departments are disappearing too. Like English universities during the modern era, German universities were tied to an idea. For Kant, it was reason. For Schiller and Humboldt, it was soul-building (Hofstetter, 2001). Hofstetter (2001) explains that for Kant the idea of the university and university life was that by "building up the mind" and honing one's reasoning ability, one could aim for a "model of perfection" (p. 26) in human thinking. Alongside these aims, anti-intellectualism began creeping into the curriculum. Hofstetter (2001) claims that Freidrich "Schelling was particularly disturbed by how the system wasted the talents of good students, pushing them into vocational tracks and giving them a contempt for Wissenschaft" (p. 55). Kant was also disturbed by students whom he referred to as careerists and warned that careerists should be kept away from real scholars. Are there scholars who are not, in some way, careerists? Or are the careerists the functionaries, those obsessed with committee meetings? Are the careerists the professors who say "I can't write because I'm too busy chairing committees."

Bill Readings (1996) sums up the modernist period by saying that "the modern university has had three ideas: the Kantian concept of reason, the Humboldtian idea of culture, and now the techno-bureaucratic notion of excellence" (p. 14). A grand generalization, but nonetheless I find Readings' summation helpful. Readings' argument is well known. The university is now in "ruins," there is no longer an idea to which the university is tied. The idea of excellence is empty, meaningless. Readings asks some interesting questions about university life and the life of intellectuals within the "ruins." Readings (1996) says "We have to recognize that the university is a ruined institution; while thinking what it means to dwell in those ruins without recourse to romantic nostalgia" (p. 169). Since we cannot return to the Kantian cry for reason and duty, or the Romantic notion of soul-building, what do we do? What is the use of the university and what are scholars' responsibilities within the ruins?

What do we do inside the halls of U.S. Steel U? or The University of Standard Oil? What kind of work is valued inside Enron U.? It seems that one cannot redeem the university. My aim is not redemption, my aim is to try to figure out the scholarly life within an institution that I do not understand. Ah—but it is our home. I cannot think of any other place I'd rather be! Many have a love-hate relation with the university. The dystopic university is the site of both thanatos and eros. It is the site of groundlessness, it is the site of a huge gaping IT, a whirling unconscious that pulls us into our future scholarly lives. To be not-at-home-inside the (un)home of the university—this is the postmodern condition, the post–9-11 situation. This is the schizophrenic motion we must psychically manage in the madhouse that is the university. David Jardine (1998) suggests that attunement to lived experience leaves us where we already are. Thus, in the case of universities, scholars are always already *thrown* into institutional life. After all is said and done, we are where we are thrown. And at present we are thrown into the madhouse that is academe. We must live the dystopic life. Here paranoia rules. One must always already be paranoid. Yet out of paranoia springs interesting work that is perhaps hard to swallow, but isn't that what cutting-edge scholarship is about. It should get your hackles up, it should give you the jim jams. The university is also a site of reparation. It is both paranoid and reparative simultaneously. It is both greedy and loving; both destructive and life-giving. It is a mess. It is life and death at once. Reflecting on scholarly life within the academy, the intellectual must grapple with the paradox that is the institution. Scholarly work is at loggerheads with the aim of university, which is *no aim* at all. What kind of an aim is an anti-intellectual one? What do we do in the Toyota School, as Philip Wexler (1996) aptly puts it. Scholars live a schizophrenic existence. Doing scholarly writing is a threat and the institution wants none of it—really. It wants money and prestige but it does not want scholars. It wants Toyota salesmen.

Dystopic University as Schizophrenic

Richard Miller (1998) says that "There are, of course, very sound reasons for seeing the world of higher education as a jumble of meaningless contradictions that can never be changed or understood" (p. 3). The dystopic university is indeed one of meaningless contradictions! But it is more psychological than just contradictory. Scholars are not the masters of their own houses within the dystopic university. The university acts as a giant superego which sadistically punishes scholars for doing their work, especially if it is theoretical and does not pull in funds. Scholars are the whipping posts for administrators. To psychically deal with sado-masochism is draining. If one tries to fight the sadistic master, the sadism comes back twofold. It is at this crazy site that one works as one does. Readings (1996) suggests that the best we can do at this historical juncture is to realize that "The university is where thought takes place beside thought . . . The university's ruins offers us an institution in which the incomplete and interminable nature of the pedagogies . . . can remind us that 'thinking together' is a dissensual process" (p. 192). Like Readings, Ewa Ziarek (2001) calls for embracing an ethic of dissensus at the site of intellectual work. She suggests we should adopt this ethical sensibility if we are to pay attention to the other, to hear what the other has to say. This is Dewey's democratic progressive project. The other who is installed within the halls of academe, risks erasure by doing work that is considered different.

University as Site of Otherness

A helpful way to think about the university is that it is a site of Otherness. It creates Otherness by its very project, by its very aim, its aim of no aim. The institution, paradoxically, creates a space of alienation for scholars who are not tied to the market. It creates a schizoid existence. For those of us who are on the left and who feel schizoid tendencies, Ziarek (2001) is instructive. She draws on the work of Levinas and suggests that we might begin to "articulate what Emmanuel Levinas calls an 'anarchic obligation,' one that signifies a nonappropriative relation to the other" (p. 2). Scholars on the left might begin to think otherwise, by relating to one another and the larger community more sensitively, by allowing one another to think differently. An anarchic obligation is one that does not follow Kantian principles, or utilitarian prescriptives, but follows aporetic intuitions that open out toward the other, that open out toward the thinking-other of the other-of-thought.

Orpheus diss-ending. In the dystopic university, scholars must go underground but yet, still remain above ground in the groundlessness that is academe. Scholars' call is to do the work of Thoth or Hermes while living on the para-site of the militarily- excellent university. These are the signs of a troubled time. This is a postmodern nightmare—but it is the only possible existence for the scholar on the left. The scholar housed in the dystopic university does the dance of the schizophrenic. We are neither here nor there, but firmly placed and yet out of place. Scholars on the left are dissenters.Yet, under the sign of excellence, (which functions as omen and as symptom) dissensus and dissent are not valued because these signi-fires could get us fired. Dissent could slip into something more serious. Thoth might undermine the notion of excellence itself. Hermes might trick.

Derrida (1992) warns that scholars must be on the look-out for language like "consensus" because it signals a supposed "transparency" (pp. 54–55). Consensus is the signification of the same. Why must scholars think alike? Are we really on the same page? Even if we think we are on the same page, I think we are not. But faculty members could be afraid to voice dissenting opinions, especially during times of war. Derrida (1992) remarks that talk of consensus "tends to impose a model of language that is supposedly favorable to this communication. Claiming to speak in the name of intelligibility, good sense, common sense . . . this discourse tends, by means of these very things, and as if naturally, to discredit anything that complicates . . . It tends to suspect or repress anything that bends, overdetermines, or even questions" (pp. 54–55). To come to consensus means to work in conjunction with the larger corporate mentality of the university. This is the call of administrators. But scholars worth their salt know that intellectual fields are not built by consensus. Fields are built from schizophrenic wanderings. Fields are built from fierce, emotional disagreements.

The notion of dissensus threatens because like the very act of writing, it signals an excess, an interminable delay, a confusion. Does dissensus mean dissent? Does dissent signal anarchy? Or does dissensus simply mean disagreement? Can we disagree with the aim of the university? What is the aim anyway? *No* aim. To whom would one address one's disagreement? The public, a public, the president, a chair, a fellow faculty member? How does a disagreement change the institution? Do institutions get changed by dissent?

Richard Miller (1998) points out that "the historical inertia of the institution and its practices will ensure that even . . . modest changes will encounter a general, if low-level, resistance" (p. 19). The institution is slow to change because change is not welcomed. The university is inherently conservative in nature. That is, the university is supposed to uphold the tradition, while holding off the new or questionable. Of course, this throws into question the entire notion of academic freedom. Is there any? I suppose this question depends upon which institution one is housed in. Joan Scott (1996) remarks that "academic freedom lives in the ethical space between an ideal of the autonomous pursuit of understanding and the specific historical, institutional, and political realities that limit such practices" (p. 177). Dissenting subject-positions are dangerous. Dissenting departments, that is, departments that are viewed by the administration as useless, not practical, are the first to be abolished. Christopher Newfield (1998) says "From my vantage point, the most immediate casualties [of retrenchment] are dreams of the new—new programs, new disciplines, new combinations" (p. 81). How new is new? Is Dewey still new? Is the reconceptualization still new? New programs, new disciplines beware.

Disturbingly, Dewey's progressive dream *still* struggles to survive in the university, a place without a center, a university without an idea, a place that, as Peggy Kamuf (1997) puts it, "founders where it is founded" (p. 55). Reconceptualized curriculum scholarship is still viewed with suspicion. Derridean scholar Simon Wortham (1999) suggests that the university is in a state of confusion, paradox, and disequilibrium. Wortham remarks,

> This imbalance we have linked to an insoluble disorientation between left and right in a university uncertain as to its ground. All sorts of leverages that occur within the university (and which shape its institutions generally: its critical orthodoxies and counter-orthodoxies; its formations of disciplinary and interdisciplinary fields; its modes and discourses of publication, etc.) are, I would support, undertaken precisely by means of intractable confusions between left and right. (p. 8)

Wortham suggests that instead of trying to resist forces on the right, one must acknowledge the "disorientation" (p. 9) that is symptomatic of the foundering. But unlike anarchists, who would like to overthrow institutions, Wortham and other poststructuralists suggest that one live in the chaos with a sense of responsibility. But this is not enough. I suggest that one be aware of the psychic damage done by university life. I suggest that the notion of the dystopic university is one that must include discussions of the psyche. To have a sense of "responsibility" is nice, but what does that really mean when one is constantly pulling one's hair out like Ezekiel at the madhouse that is the university. How to have responsibility when one is paranoid all the time about what knowledge is of most worth and what counts and what does not count? Is this just accounting? Yes, it is all about numbers and not quality. But the administration will tell junior faculty that it is all about quality and not numbers. It is an Alice in Wonderland game. And yet—it is almost too simple—and impossible—to say "the hell with it!" Well, I suppose if you were rich you could say the hell with it!! But scholars need homes, we need communities, even if we cannot speak to each other. Still, for psychological reasons we need each other. Scholars need community, even if that community is contentious. Wortham stresses, "The institution . . . is based, founded, on a monstrosity . . . which nevertheless need not—indeed cannot—simply be negatively marked" (p. 31). He suggests that the university is a "monstrosity" (p. 42) and will always already be in a state of "dis-re-pair" (p. 42). This does not mean, however, that intellectuals housed within monstrous institutional sites do nothing! This is not a call for nihilism. Derrida suggests that scholars take responsibility for trying to free up the institution by engagement. Derrida (1995b) says,

> If an "ideologue" or "intellectual" does not attempt to transform effectively the cultural, academic, or editorial apparatuses in which he works, whether he is sleepily installed there or still claims to be free to "wander" within it . . . than he is always in the process of maintaining the good working order of the most sinister of machines. Sometimes this is accompanied, as a confirmation, by an overt, disdainful, and moralizing lack of interest in matters having to do with teaching and publishing. (p. 63)

Kafka-esque characters are indeed lazily installed within the institution and they are the ones who claim that they do not have time to publish because they are so busy chairing committee meetings, or argue that writing does not matter. Ah, here are the sophists. Through our everyday work and life within the halls of academe, scholars "live on" as Derrida might say, toward a more "responsible response." And this responsible response begins in daily study—and attention to psychic well being or injury—and engaging in the hard and threatening work of writing. Kafka's characters can carry on if they wish, but we thinkers on the left will continue the hard work of theorizing curriculum.

This is a response not of redemption, but perhaps of (post) reconceptualization. Dewey's dream "lives on" within the discipline of curriculum studies, beside the para-sites of excellence, the call toward the practical and the "cultural take over" of consensus. A schizophrenic existence is one that makes little sense. And certainly life within the university makes little sense. Intellectual work within the university is beset with paradoxes and crises. Faculty scramble to get a grip. But there is nothing to grip but the slippery slope of change. The institution is a living creature. And like all living creatures it eludes our grasp. Scholars live in a dystopic university where Orwell's prophecy speaks.

A Phenomenology of the Dystopic University

The dystopic university is a place where there is no longer any grounding or purpose for what we do. We are a lost generation of scholars, lost in the dystopic university, lost in our increasingly anti-intellectual culture, lost in the cyberspace of gadgets. We are the generation of cell phones, Columbine, 9-11, high speed Internet, highway shootings, gameboys, nuclear proliferation, the lie of weapons of mass destruction, biological terrorism. We are lost in our cybernet culture where anyone can learn to build a bomb on the Internet, where kids kill each other over a pair of tennis shoes, where human life is no longer valued, where beheadings can be seen on TV and on the Internet. Our work as scholars is not appreciated by the very institution that houses us because the institution has lost its way. The institution squeezes the life out of us like a copperhead. We are intellectuals who work within a dystopic space. Our scholarship cannot not be dystopic. But we must reflect lived experience dystopically and honestly. To live inside a dystopic space is to be lost, is to be necessarily lost. To live inside a dystopic space is to feel a foundering, a groundlessness, to feel lost in a void, a chiasma, a whirlpool of hithering. We are a lost generation of negative reverie, of black holes, of miasmas, of trepidation in still waters. We fear for our childrens' future. We fear for the future. Will there even be a future? Will our political "leaders" just blow the planet to hell after all? The dystopic university is a minor 5th and a suspended 7th. It is a chord without resolution, forever suspended in time, it is out of time and out of joint. How can one do scholarship thinking these things? One must work dystopically if one is to be in touch with our current political nightmare. The dystopic university is a nightmare and a dream of possibility. The dystopic university is schizophrenic. It is a place where we must fill the void by writing, writing, writing. Filling up the blank spaces like Artaud and Schreber. Filling up the void by pages and pages. The dystopic university is a place that has not gone beyond the paranoid-schizoid position. The dystopic university is a site where objects and subjects are symbiotic, where people are objects. You are a radio-head, and I am an elephant in the flat. There is no warmth inside the dystopic university; it is as if scholars are born too soon or are traumatized by their birth as scholars. Scholars suffer from psychic wounds. The dystopic university is the not good enough mother, the schizoid mother. The dystopic university is the law of the father, the sadistic superego. Scholars feel perpetually out of it, run down, dissociated, zoned out. The dystopic university is a place of whirling where there is no bottom, a place of disconnected thinking and dissociated thought. The dystopic university is an

Alice in Wonderland *whirl of jazzlike improvisation and strict classical interpretation. The dystopic university is the mother who does not love her child, it is the mother who drinks too much and takes valium and turns on the engine of the car while sitting in a closed garage. The dystopic university is the father who beats his child. The dystopic university is the Wolf Man, the Rat Man, and Dr. Paul Schreber. The dystopic university is the paranoid-schizophrenic position; it is a place of negative reverie it is a place of no-memory, it is a place where there is no past and no future because it has lost its way. The dystopic university is repressed memory, it is lost, forever lost in a post-Columbine state. The dystopic university is a place of ecstasy. Here the scholar feels wild with freedom and the passions of a Jack Kerouac* On the Road, *a Clint Eastwood* Easy Riderness. *The dystopic university is a reparative site. Here, writing has no end, writing is done in the heat of the night with great passion and fury. The dystopic university is the reparative space, the depressive position, the shining sun, the California haze, a gray Pittsburgh day. The dystopic university is the Dali Lama of one-ness and longing and the suchness of suffering. It is a place where ideas are born and friendships are made. Enemies abound as well. The dystopic university is always already a space of life unlike any other. The dystopic university is the harshness that is the midwest and it is the softness that is the ocean. Scholars are on the road in their minds while traveling nowhere. The dystopic university is where we find our company in books, our company in the company of scholars, where there are no longer any scholars and there is no company to be found. The dystopic university is confusion and ungrounding. The dystopic university has no aim. The dystopic university is a site of Otherness.*

Bibliography

Adorno, T. W. (1992). *Mahler: A musical physiognomy* (Trans., E. Jephcolt). Chicago, IL: University of Chicago Press.

———. (1995). *Negative dialectics* (Trans., E. B. Ashton). New York: Continuum.

Agassi, J. B. (1999). Preface. In J. B. Agassi (Ed.), *Martin Buber on psychology and psychotherapy: Essays, letters, and dialogue* (pp. vii–xvi). New York: Syracuse University Press.

Alexander, R. P. (1993). On the analyst's "sleep" during the psychoanalytic session. In J. S. Grotstein (Ed.), *do i dare disturb the universe? a memorial to W. R. Bion* (pp. 46–57). London: Karnac.

Amery, J. (1999). Antisemitism on the left—The respectable antisemitism. In D. C. G. Lorenz (Ed.), *Contemporary Jewish writing in Austria: An anthology* (pp. 115–130). Lincoln, NE: University of Nebraska Press.

Appignanesi, L. and J. Forrester. (2000). *Freud's women.* New York: The Other Press.

Arendt, H. (1958/1998). *The human condition.* Chicago, IL: University of Chicago Press.

———. (1978). *The life of the mind.* New York: A Harvest Book, Harcourt Brace & Company.

———. (1992). *Hanna Arendt, Karl Jaspers correspondence 1926–1969* (Trans., R. and R. Kimber). New York: A Harvest Book, Harcourt Brace & Company.

Artaud, A. (1988a). Fragments from a diary from hell (1925). In S. Sontag (Ed.), *Antonin Artaud: Selected writings* (Trans., H. Weaver) (pp. 91–96). Berkeley, CA: University of California Press.

———. (1988b). Letter to George Soulie De Morant, February 17, 1932. In S. Sontag (Ed.), *Antonin Artaud: Selected writings* (Trans., H. Weaver) (pp. 286–290). Berkeley, CA: University of California Press.

———. (1988c). Letter to Jacques Riviere. In S. Sontag (Ed.), *Antonin Artaud: Selected writings* (Trans., H. Weaver) (pp. 31–33). Berkeley, CA: University of California Press.

———.(1988d). Letter to Pierre Loeb, April 23, 1947. In S. Sontag (Ed.), *Antonin Artaud: Selected writings* (Trans., H. Weaver) (pp. 515–519). Berkeley, CA: University of California Press.

———. (1988e). Van Gough, the man suicided by society (1947). In S. Sontag (Ed.), *Antonin Artaud: Selected writings* (Trans., H. Weaver) (pp. 483–512). Berkeley, CA: University of California Press.

Artaud, A. (1988f). Letter to the legislator of the law on narcotics. In S. Sontag (Ed.), *Antonin Artaud: Selected writings* (Trans., H. Weaver) (pp. 68–72). Berkeley, CA: University of California Press.

Atwell-Vasey, W. (1998). *Nourishing words: Bridging private reading and public teaching*. Albany, NY: SUNY.

Bance, A. (1932/1995). Introduction. In Joseph Roth's *The Radetzky march* (Trans., J. Neugroschel) (ix–xxx). New York: Alfred A. Knopf.

Beckett, S. (1965). *Three novels* (Trans., P. Bowles). New York: Grove Press.

———. (1980). *Company*. New York: Grove Press.

Bell, D. (1999). Destructive narcissism and the singing detective. In P. Williams (Ed.), *Psychosis (madness)* (pp. 78–92). London: Institute of Psychoanalysis.

Beller, S. (1987). Class, culture and the Jews of Vienna, 1900. In I. Oxaal, M. Pollack, and G. Botz (Eds.), *Jews, antisemitism and culture in Vienna* (pp. 39–58). New York: Routledge & Kegan Paul.

Benjamin, W. (1994). *The correspondence of Walter Benjamin* (Trans., M. R. Jacobson and E. M. Jacobson). Chicago, IL: University of Chicago Press.

Bentolila, D. (2000). Lacan in America. In K. R. Malone and S. R. Friedlander (Eds.), *The subject of Lacan: A Lacanian reader for psychologists* (pp. 317–329). Albany, NY: SUNY.

Bergman, I. (1960). *The seventh seal* (Trans., L. Malmstrom and D. Kushner). New York: Simon & Schuster.

———. (1982). *Fanny and Alexander* (Trans., A. Blair). New York: Pantheon Books.

Berio, T. P. (2002). Mahler's Jewish parable. In K. Painter (Ed.), *Mahler and his world* (pp. 87–110). Princeton, NJ: Princeton University Press.

Bernstein, R. (1997). Pragmatism, pluralism, and the healing of wounds. In L. Menand (Ed.), *Pragmatism: A reader* (pp. 382–401). New York: Vintage.

Bettelheim, B. (1989a). *The uses of enchantment: The meaning and importance of fairy tales.* New York: Vintage.

———. (1989b). *Freud's Vienna and other essays.* New York: Vintage.

Bischof, G. (1997a). Founding myths and compartmentalized past: New literature on the construction, hibernation, and deconstruction of WWII memory in postwar Austria. In G. Bischof and A. Pelinkia (Eds.), *Austrian historical memory and national identity* (pp. 302–341). New Brunswick, NJ: Transaction Publishers.

———. (1997b). Introduction. In G. Bischof and A. Pelinka (Eds.), *Austrian historical memory and national identity* (pp. 1–19). New Brunswick, NJ: Transaction Publishers.

Bion, W. R. (1967/1993a). Development of schizophrenic thought. In W. R. Bion,*Second thoughts: Selected papers on psychoanalysis* (pp. 36–42). London: Karnac.

———. (1967/1993b). Differentiation of the psychotic from the non-psychotic personalities. In W. R. Bion, *Second thoughts: Selected papers on psychoanalysis* (pp. 43–64). London: Karnac.

———. (1967/1993c). Attacks on linking. In W. R. Bion, *Second thoughts: Selected papers on psychoanalysis* (pp. 93–109). London: Karnac.
———. (1989a). *Elements of psychoanalysis.* London: Karnac.
———. (1989b). Caesura. In W. R. Bion, *Two papers: The grid & caesura* (pp. 37–56). London: Karnac.
———. (1991a). Contributions to panel discussions: Brasilia, a new experience. In W. R. Bion, *All my sins remembered, another part of life: The other side of genius, family letters* (pp. 131–139). London: Karnac.
———. (1991b). Sao Paulo 1978. In W. R. Bion, *All my sins remembered, another part of life: The other side of genius, family letters* (pp. 141–240). London: Karnac.
———. (1991c). *Learning from experience.* London: Karnac.
———. (1993). *Attention and interpretation.* London: Karnac.
———. (1994a). 25 July 1959. In F. Bion (Ed.), *Cogitations: New extended edition* (pp. 62–68). London: Karnac.
———. (1994b). 13 September 1959. In F. Bion (Ed.), *Cogitations: New extended edition* (pp. 74–75). London: Karnac.
———. (2000). "Brasilia" (1975). In F. Bion (Ed.), *Clinical seminars and other works: Wilfred R. Bion* (pp. 3–139). London: Karnac.
Bleandonu, G. (2000). *Wilfred Bion: His life and works 1897–1979.* New York: The Other Press.
Bleich, D. (1999). The living text: Literary pedagogy and Jewish identity. In H. S. Shapiro (Ed.), *Strangers in the land: Pedagogy, modernity, and Jewish identity* (pp. 109–132). New York: Peter Lang.
Bleuler, E. (1911/1950). *Dementia praecox or the group of schizophrenias* (Trans., J. Zinkin). New York: International Universities Press.
Bloch, D. I. (1997). *The book of Ezekiel chapters 1–24.* Grand Rapids, MI: William B. Eerdmans Publishing Company.
Block, A. A. (1997). *I'm only bleeding: Education as the practice of violence against children.* New York: Peter Lang.
———. (1999). On singing the lord's song in a foreign land. In H. S. Shapiro (Ed.), *Strangers in the land: Pedagogy, modernity, and Jewish identity* (pp. 153–188). New York: Peter Lang.
Blum, H. P. (1999). Freud at the crossing of the millennia. *The Journal of the American Psychoanalytic Association*, 47 (4), pp. 1027–1035.
Bollas, C. (1989). *Forces of destiny: Psychoanalysis and human idiom.* North Road, London: Free Association Books.
———. (1997). "Christopher Bollas." In A. Molino (Ed.), *Elaborate selves: Reflections and reveries of Christopher Bollas, Michael Eigen, Poly Young-Eisendrath, Samuel & Evelyn Laeuchli, and Marie Coleman Nelson* (pp. 11–60). New York: The Hawthorn Press.
———. (1999). *The mystery of things.* New York: Routledge.
Botstein, L. (2002). Whose Gustav Mahler? Reception, interpretation, and history. In K. Painter (Ed.), *Mahler and his world* (pp. 1–34). Princeton, NJ: Princeton University Press.

Boyarin, J. and D. Boyarin. (1997). *Jews and other differences: The new Jewish cultural studies.* Minneapolis, MN: University of Minnesota Press.

———. (2002). *Powers of diaspora: Two essays on the relevance of Jewish culture.* Minneapolis, MN: University of Minnesota Press.

Breger, L. (2000). *Freud: Darkness in the midst of vision.* New York: John Wiley & Sons.

Britzman, D. (1998). Is there a queer pedagogy? Or, stop reading straight. In W. F. Pinar (Ed.), *Curriculum toward new identities* (pp. 211–231). New York: Garland.

———. (2003). *After-education: Anna Freud, Melanie Klein, and psychoanalytic histories of learning.* Albany, NY: SUNY.

Brosio, R. (1998). End of the millennium: Capitalisms's dynamism, civic crises, and corresponding consequences for education. In H. S. Shapiro and D. E. Purpel (Eds.), *Critical social issues in American education: Transformation in a postmodern world* (pp. 27–44). Mahwah, NJ: Lawrence Erlbaum and Associates Publishers.

Browning, C. (1998). *Ordinary men: Reserve police battalion 101 and the final solution in Poland.* Cambridge, MA: Harvard University Press.

Buber, M. (1947/2002). *Between man and man* (Trans., R. G. Smith). New York: Routledge.

———. (1948/1976). *Israel and the world: Essays in a time of crisis.* New York: Syracuse University Press.

———. (1958/1986). *I and thou* (Trans., R. G. Smith). New York: Collier Books; Macmillan Publishing Company.

———. (1992a). The nature of man. In S. N. Eisenstadt (Ed.,), *On intersubjectivity and cultural creativity.* (pp. 29–41). Chicago, IL: University of Chicago Press.

———. (1992b). The social dimensions of man. In S. N. Eisenstadt (Ed.,), *On intersubjectivity and cultural creativity.* (pp. 57–80). Chicago, IL: University of Chicago Press.

Carlson, D. (1998). Education as a political issue (I). In H. S. Shapiro and D. E. Purpel (Eds.), *Critical social issues in American education: Transformation in a postmodern world* (pp. 279–288). Mahwah, NJ: Lawrence Erlbaum and Associates Publishers.

Campbell, J. (1972). *Myths to live by: How we re-create ancient legends in our daily lives to release human potential.* New York: Penguin Compass.

Cohen, G. B. (1996). *Education and middle-class society in imperial Austria 1848–1918.* West Lafayette, IN: Purdue University Press.

Coles, R. (1992). *Anna Freud: The dream of psychoanalysis.* Cambridge, MA: Perseus Publishing.

Coltart, N. (2000). *Slouching toward Bethlehem.* New York: The Other Press.

Crankshaw, E. (1963). *The fall of the house of Habsburg.* New York: Penguin.

Cushman, P. (1995). *Constructing the self, constructing America: A cultural history of psychotherapy.* Reading, MA: Addison-Wesley.

Daspit, T. and J. Weaver. (Eds.) (2000). *Popular culture and critical pedagogy: Reading, constructing and connecting.* New York: Garland.

de Tocqueville, A. (2000). *Democracy in America* (Trans. G. Lawrence). (Ed.) J. P. Mayer. New York: Perennial Classics.

Deleuze, G. and F. Guattari. (1987). *A thousand plateaus: Capitalism and schizophrenia* (Trans., B. Massumi). Minneapolis, MN: University of Minnesota Press.

———. (2000). *Anti-Oedipus:Capitalism and schizophrenia* (Trans., R. Hurley, M. Seems, and H. R. Lane). Minneapolis, MN: University of Minnesota Press.

Derrida, J. (1976). *Of grammatology* (Trans., G. C. Spivak). Baltimore, MD: Johns Hopkins University Press.

———. (1978). *Writing and difference* (Trans., A. Bass). Chicago, IL: The University of Chicago Press.

———. (1981). *Dissemination* (Trans., B. Johnson). Chicago, IL: University of Chicago Press.

———. (1987). *The postcard: From Socrates to Freud and beyond* (Trans., A. Bass). Chicago, IL: University of Chicago Press.

———. (1992). *The other heading: Reflections on today's Europe* (Trans., P. Brault and M. B. Nass). Bloomington, IN: Indiana University Press.

———. (1994). *Specters of Marx: The state of debt, the work of mourning and the new international* (Trans., P. Kamuf). New York: Routledge.

———. (1998a). *Resistances of psychoanalysis* (Trans., P. Kamuf, P. A. Brault, and M. Nass). Stanford, CA: Stanford University Press.

———. (1998b). To unsense the subjectile. In J. Derrida and P. Thevenin, *The secret art of Antonin Artaud* (pp. 59–157) (Trans., M. A. Caws). Cambridge, MA: MIT Press.

———. (2000). Where a teaching body begins and how it ends. In P. Trifonas (Ed.), *Revolutionary pedagogies: Cultural politics, instituting education, and the discourse of theory* (pp. 83–112). New York: Routledge.

———. (2001). "Negotiations." In E. Rottenberg (Ed./Trans.,) *Jacques Derrida negotiations: Interventions and interviews 1971–2001* (pp. 11–40). Stanford, CA: Stanford University Press.

———. (2002). Faith and knowledge: The two sources of religion at the limits of reason alone (Trans., S. Webber). In G. Anidjar (Ed.), *Acts of religion Jacques Derrida* (pp. 230–298). New York: Routledge.

———. (2002c). *The university without condition.* In P. Kamuf (Ed./Trans.,) *Without Alibi* (pp. 202–237). Stanford, CA: Stanford University Press.

———. (2002d). Hospitality (Trans, G. Anidjar). In G. Anidjar (Ed.), *Acts of religion Jacques Derrida* (pp. 358–420). New York: Routledge.

———. (2002e). *Who's afraid of philosophy? Right to philosophy I* (Trans., J. Plug). Stanford, CA: Stanford University Press.

Derrida, J. and B. Stiegler. (2002a). *Echographies of television.* Malden, MA: Polity Press/ Blackwell.

DeSalvo, L. (1998). Advice to aspiring educational biographers. In C. Kridel (Ed.), *Writing educational biography: Explorations in qualitative research* (pp. 269–271). New York: Garland.

Dewey, J. (1997/1917). From the need for a recovery of philosophy. In L. Menand (Ed.), *Pragmatism: A reader* (pp. 219–232). New York: Vintage.

Dick, K. and Amy Ziering Kofman. (2003). "Derrida." Zeitgeist Video, Z1047.
Doll, M. A. (1988). *Beckett and myth: An archetypal approach.* New York: Syracuse University Press.
———. (1995). *To the lighthouse and back: Writings on teaching and living.* New York: Peter Lang.
———. (2000). *Like letters in running water: A mythopoetics of curriculum.* Mahwah, NJ: Lawrence Erlbaum and Associates Publishers.
———. (2004). *Blathering in the void: A talk on talk in Beckett.* Public Lecture at the Afifi Amphitheater, April 29, 2004, Savannah, Georgia.
Dreyfus, H. (1954/1987). Foreword. In M. Foucault (Ed.), *Mental illness and psychology* (pp. vii–xliii). Berkeley, CA: University of California Press.
Durrell, L. (1961). Introduction. In G. Groddeck's *The book of the it* (v–xxiv). New York: Vintage.
Dyer, R. (1983). *Her father's daughter: The work of Anna Freud.* New York: Jason Aronson.
Eigen, M. (1993). *The psychotic core.* Northvale, NJ: Jason Aronson.
———. (1996). *Psychic deadness.* Northvale, NJ: Jason Aronson.
———. (1997). "Michael Eigen." In A. Molino (Ed.), *Elaborate selves: Reflections and reveries of Christopher Bollas, Michael Eigen, Polly Young-Eisendrath, Samuel and Evelyn Laeuchli and Marie Coleman Nelson* (pp. 105–141). New York: The Hawthorn Press.
———. (1998). *The psychoanalytic mystic.* London: Free Association Books.
———. (2001a). *Ecstasy.* Middletown, CT: Wesleyan University Press.
———. (2001b). *Toxic nourishment.* London: Karnac.
———. (2004). *The sensitive self.* Middletown, CT: Wesleyan University Press.
Elon, A. (2002).*The pity of it all: A portrait of the German-Jewish epoch 1743–1933.* New York: Picador, Henry Holt and Company.
Estonis, S. (2002). S. Estonis. The Autobiographies. In J. Sandler (Ed.), *Awakening lives: Autobiographies of Jewish youth in Poland before the Holocaust* (pp. 3–19). New Haven, CT: Yale University Press.
Fairbairn, R. (1954). *An object-relations theory of personality.* New York: Basic Books.
Federn, E. (1990). *Witnessing psychoanalysis: From Vienna back to Vienna via Buchenwald and the USA.* London: Karnac.
Fenichel, O. (1953a). Identification. In H. Fenichel and D. Rapaport (Eds.), *The collected papers of Otto Fenichel: First Series* (pp. 97–112). New York: W.W. Norton.
———. (1953b). The pregenital antecedents of the Oedipus complex. In H. Fenichel and D. Rapaport (Eds.), *The collected papers of Otto Fenichel: First Series* (pp. 181–203). New York: W.W. Norton.
Feuerverger, G. (2001). My Yiddish voice. In M. Morris, and J. Weaver (Eds.), *Difficult memories: Talk in a (post) Holocaust era* (pp. 13–24). New York: Peter Lang.

Ficowski, J. (1999). "Interview in Warsaw, July 3, 1993." Dorota Glowacaka and Jerzy Ficowski. In C. Z. Prokopczyk (Ed.), *Bruno Schulz: New documents and interpretations* (pp. 55–69). New York: Peter Lang.

———. (2003). *Regions of the great heresy: Bruno Schulz a biographical portrait* (Trans., T. Robertson). New York: W.W. Norton.

Finkelstein, N. (1996). "The master of turning" Walter Benjamin, Gershom Scholem, Harold Bloom, and the writing of a Jewish life. In J. Rubin-Dorsky and S. F. Fishkin (Eds.), *People of the book: Thirty scholars reflect on their Jewish identity* (pp. 415–426). Madison, WI: University of Wisconsin Press.

Floros, C. (2000). *Gustav Mahler: The symphonies* (Trans., V. and J. Wicker). Portland, OR: Amadeus Press.

Foucault, M. (1954/1987). *Mental illness and psychology* (Trans., A. Sheridan). Berkeley, CA: University of California Press.

———. (1965/1988). *Madness and civilization: A history of insanity in the age of reason* (Trans., R. Howard). New York: Vintage.

Freundlich, E. (1999). Excerpt from the soul bird. In D. C. G. Lorenz (Ed.), *Contemporary Jewish writing in Austria: An anthology* (pp. 5–21). Lincoln, NE: University of Nebraska Press.

Freud, A. (1926/2000). Letter # 1068 to Sandor Ferenczi. In E. Falzeder and E. Brabant (Eds.,) in collaboration P. G. Deutsch. *The correspondence of Sigmund Freud and Sandor Ferenczi* (Trans., P. Hoffer) (p. 269). Cambridge, MA: Belknap Press of Harvard University Press.

———. (1927/2000/). Letter # 1089 to Sandor Ferenczi. In E. Falzeder and E. Brabant (Eds.,) in collaboration P.G. Deutsch *The correspondence of Sigmund Freud and Sandor Ferenczi* (Trans., P. Hoffer) (pp. 297–298). Cambridge, MA: Belknap Press of Harvard University Press.

———. (1935/1989). *An autobiographical study.* (Trans., J. Strachey). The Standard Edition. New York: W.W. Norton.

———. (1960a). *Jokes and their relation to the unconscious* (Trans., J. Strachey). The Standard Edition. New York: W.W. Norton.

———. (1960b). Letter # 55 to Martha Bernays 1885. In E. L. Freud (Ed.), *Letters of Sigmund Freud* (Trans., T. and J. Stern) (pp. 131–132). New York: Basic Books.

———. (1960c). Letter # 94 to Martha Bernays 1886. In E. L. Freud (Ed.), *Letters of Sigmund Freud* (Trans., T. and J. Stern) (p. 200). New York: Basic Books.

———. (1960d). Letter # 126 to C. G. Jung 1907. In E. L. Freud (Ed.), *Letters of Sigmund Freud* (Trans., T. and J. Stern) (p. 256). New York: Basic Books.

———. (1960e). Letter # 219 to Enrico Morselli 1926. In E. L. Freud (Ed.), *Letters of Sigmund Freud* (Trans., T. and J. Stern) (p. 365). New York: Basic Books.

———. (1960f). Letter # 285 to Arnold Zweig 1936. In E. L. Freud (Ed.), *Letters of Sigmund Freud* (Trans., T. and J. Stern) (pp. 430–438). New York: Basic Books.

Freud, A. (1963). *Psycho-analysis for teachers and parents* (Trans., B. Low). New York: W.W. Norton.

———. (1967). *Moses and monotheism* (Trans., K. Jones). New York: Vintage Press.

Freud, A. (1969a). Assessment of pathology in childhood. In *The writings of Anna Freud, volume v: Research at the Hampstead child-therapy clinic and other papers 1956–1965* (pp. 26–59). New York: International Universities Press.

———. (1969b). Child observation and prediction of development: A memorial lecture in honor of Ernst Kris. In *The writings of Anna Freud, volume v: Research at the Hampstead child-therapy clinic and other papers 1956–1965* (pp. 102–135). New York: International Universities Press.

———. (1969c). The child guidance clinic as a center of prophylaxis and enlightenment. In *The writings of Anna Freud, volume v: Research at the Hampstead child-therapy clinic and other papers 1956–1965* (pp. 281–300). New York: International Universities Press.

———. (1995). Address to the society of B'nai B'rith. In S. Gilman (Ed.), *Sigmund Freud: Psychological writings and letters* (pp. 266–268). New York: Continuum.

———. (2004). The reading cure: Books and lifetime companions. *American Imago: Studies in Psychoanalysis and Culture*. 61 (1), 77–87.

Friedlander, S. (1999). "Europe's inner demons": The "other" as threat in early thirteenth-century European culture. In R. S. Wistrich (Ed.), *Demonizing the other: Antisemitism, racism and xenophobia* (pp. 210–222). Amsterdam, The Netherlands: Harwood Academic Publishers.

Friedman, M. (1965/1989). Introductory essay. In M. Freidman (Ed.), Martin Buber's *The knowledge of man: Selected essays* (Trans., M. Friedman and R. G. Smith) (viii–xxii). Atlantic Highlands, NJ: Humanities Press International, Inc.

———. (1991). *Encounter on the narrow ridge: A life of Martin Buber*. New York: Paragon House.

Gadamer, H-G. (1995). *Truth and method*. Second, Revised Edition (Trans., J. Weinsheimer and D. G. Marshall). New York: Continuum.

Gay, P. (1998). *Freud: A life for our time*. New York: W.W. Norton.

———. (2002). *Schnitzler's century: The making of middle-class culture 1815–1914*. New York: W.W. Norton.

Gerber, D. A. (1996). Visiting bubbe & zeyde: How I learned about American pluralism before writing about it. In J. Rubin-Dorsky and S. F. Fishkin (Eds.), *People of the book: Thirty scholars reflect on their Jewish identity* (pp. 117–134). Madison, WI: University of Wisconsin Press.

Gilman, S. (1982). *Seeing the insane*. Lincoln, NE: University of Nebraska Press.

———. (1993). *The case of Sigmund Freud: Medicine and Identity at the Fin de Siecle*. Baltimore, MD: Johns Hopkins University Press.

———. (1994). *Disease and representation: Images of illness from madness to AIDS.* Ithaca, NY: Cornell University Press.

———. (1995a). *Franz Kafka, The Jewish patient.* New York: Routledge.

———. (1995b). Introduction. In S. Gilman (Ed.), *Sigmund Freud: Psychological writings and letters* (pp. vii–xli). New York: Continuum.

———. (1997). *Smart Jews: The construction of Jewish superior intelligence.* Lincoln, NE: University of Nebraska Press.

———. (1998). *Love+marriage = death: And other essays on representing difference.* Stanford, CA: Stanford University Press.

———. (1999). Introduction: The frontier as a model for Jewish history. In S. Gilman and M. Shain (Eds.), *Jewries at the frontier: Accommodation, identity, conflict* (pp. 1–25). Urbana, IL: University of Illinois Press.

Giroux, H., D. Shumway, P. Smith, and J. Sosnoski. (1995). The need for cultural studies: Resisting intellectuals and oppositional public spheres. In J. Munns and G. Rajan (Eds.), *A cultural studies reader: History, theory, practice* (pp. 647–658). New York: Longman.

Glowacka, D. (1999). Sublime trash and the simulacrum: Bruno Schulz in the postmodern neighborhood. In C. Z. Prokopczyk (Ed.), *Bruno Schulz: New documents and interpretations* (pp. 79–121). New York: Peter Lang.

Goldhagen, D. (1997). *Hitler's willing executioners: Ordinary Germans and the Holocaust.* New York: Vintage.

Goren, A. A. (1994). Ben Halpern: At home in exile. In C. Kessner (Ed.), *The "other" New York Jewish intellectuals* (pp. 71–100). New York: New York University Press.

Green, A. (1999). *The work of the negative* (Trans., A. Weller). New York: Free Association Books.

Greene, M. (1973). *Teacher as stranger.* Belmont, CA: Wadworth.

Gresser, M. (1994). *Dual allegiance: Freud as a modern Jew.* New York: SUNY.

Groddeck, G. (1961). *The book of the it.* New York: Vintage.

———. (1977/1917). *The meaning of illness: Selected psychoanalytic writings* (Trans., G. Mander). New York: International Universities Press.

Grosskurth, P. (1986). *Melanie Klein and her work.* New York: Alfred A. Knopf.

Grotstein, J. S. (1993). Wilfred R. Bion: The man, the psychoanalyst, the mystic: A perspective on his life and work. In J. S. Grotstein (Ed.), *do i dare disturb the universe? a memorial to W. R. Bion* (pp. 1–35). London: Karnac.

———. (2000). *Who is the dreamer who dreams the dream? A study of psychic presences.* Hillsdale, NJ: The Analytic Press.

Guattari, F. (1995). *Chaosmosis: An ethico-aesthetic paradigm.* Bloomington, IN: Indiana University Press.

Gubar, S. (1996). Eating the bread of affliction: Judaism and feminist criticism. In J. Rubin-Dorsky and S. F. Fishkin (Eds.), *People of the book:*

Thirty scholars reflect on their Jewish identity (pp. 15–36). Madison, WI: University of Wisconsin Press.

Hamann, B. (1999). *Hitler's Vienna: A dictator's apprenticeship* (Trans., T. Thornton). New York: Oxford University Press.

Hanák, P. (1998). *The garden and the workshop: Essays on the cultural history of Vienna and Budapest.* Princeton, NJ: Princeton University Press.

Handlebauer, B. (1999). The influence of Austrian émigrés on the development and expansion of psychoanalysis in the United States after 1945. In D. F. Good and R. Wodak (Eds.), *From world war to Waldheim: Culture and politics in Austria and the United States* (pp. 109–137). New York: Berghahn Books.

Haraway, D. (1989). *Primate visions: Gender, race and nature in the world of modern science.* New York: Routledge.

Heidegger, M. (1927/1962). *Being and time* (Trans., J. Macquarrie and E. Robinson). New York: Harper and Row.

Heller, P. (1999). Bruno Schulz, story teller: A reader's response. In C. Z. Prokopczyk (Ed.), *Bruno Schulz: New documents and interpretations* (pp. 163–173). New York: Peter Lang.

Heschel, A. J. (1969). *The prophets: An introduction.* New York: Harpers.

Hilberg, R. (1961/1985). *The destruction of the European Jews volume 1, revised and definitive edition.* New York: Holms and Meier.

Hobsbawm, E. (1989a). *The age of empire 1875–1914.* New York: Vintage.

———. (1989b). The end of empires. In P. Pabisch (Ed.), *From Wilson to Waldheim: Proceedings of a workshop on Austrian-American relations 1917–1987* (pp. 12–16). Riverside, CA: Adriane Press.

Hofmann, M. (2002). Introduction. In *The collected stories of Joseph Roth* (Trans., M. Hofmann) (pp. 9–14). New York: W.W. Norton.

Hofstadter, R. (1962). *Anti-intellectualism in American life.* New York: Vintage.

Hofstetter, M. (2001). *The romantic idea of a university: England and Germany, 1770–1850.* New York: Palgrave.

Huebner, D. (2000a). Curriculum as concern for man's temporality. In W. F. Pinar (Ed.), *Curriculum studies: The reconceptualization* (pp. 237–249). Troy, NY: Educator's International Press, Inc.

———. (2000b). Poetry and power: The politics of curricular development. In W. F. Pinar (Ed.), *Curriculum studies: The reconceptualization* (pp. 271–280). Troy, NY: Educator's International Press, Inc.

Jackson, M. and P. Williams. (1994). *Unimaginable storms: A search for meaning in psychosis.* London: Karnac.

Jacobson, L. (1999). Whither psychoanalysis? Thirteen theses in justification and outline of a project. In R. M. Prince (Ed.), *The death of psychoanalysis: Murder? Suicide? or rumor greatly exaggerated?* (pp. 207–227). Northvale, NJ: Jason Aronson.

Jacoby, R. (1986). *The repression of psychoanalysis: Otto Fenichel and the political Freudians.* Chicago, IL: University of Chicago Press.

———. (1987). *The last intellectuals: American culture in the age of academe.* New York: Basic Books.

———. (1997). *Social amnesia: A critique of contemporary psychology.* New Brunswick: Transaction Publishers.

James, W. (1951). The moral philosopher and the moral life. In M. H. Fisch (Ed.), *Classic American philosophers: Pierce, James, Royce, Santayana, Dewey, Whitehead* (pp. 165–180). New Jersey: Prentice-Hall.

———. (1997/1907). What pragmatism means. In L. Menand (Ed.), *Pragmatism: A reader* (pp. 93–111). New York: Vintage.

Janik, A. and S. Toulmin (1996). *Wittgenstein's Vienna.* Chicago, IL: Ivan R. Dee Publishers.

Jardine, D. (1998). *To dwell with a boundless heart: Essays in curriculum theory, hermeneutics, and the ecological imagination.* New York: Peter Lang.

———. (2002). "The way to god and a true life": A spiel on Martin Heidegger, Edmund Husserl and the necessity of interpretation to a livable pedagogy. In M. Morris and J. Weaver (Eds.), *Difficult memories: Talk in a (post) Holocaust era* (pp. 209–226). New York: Peter Lang.

Jardine, D., P. Clifford, and S. Friesen. (2003). *Back to the basics of teaching and learning: Thinking the world together.* Mahwah, NJ: Lawrence Erlbaum and Associates Publishers.

Jaspers, K. (1932/1971). *Philosophy volume 3* (Trans., E. B. Ashton). Chicago, IL: University of Chicago Press.

Jones, E. (1959). *The life and work of Sigmund Freud volume 1 the formative years and the great discoveries 1856–1900.* New York: Basic Books.

Jonsson, S. (2000). *Robert Musil and the history of modern identity: Subject without nation.* Durham, NC: Duke University Press.

Johnston, W. M. (1972). *The Austrian mind: An intellectual and social history 1848–1938.* Berkeley, CA: University of California Press.

Joseph, B. (2000). Projective identification—some clinical aspects. In E. B. Spillus (Ed.), *Melanie Klein today: Developments in theory and practice: Volume 1: Mainly theory* (pp. 138–150). New York: Routledge.

Jung, K. (1963). *Memories, dreams, reflections.* In A. Jaffe (Ed.). New York: Pantheon Books.

Kafka, F. (1915/1972). *The metamorphosis.* New York: Bantam Books.

———. (1958). *The castle.* New York: Schocken Books.

Kammen, M. (1980). *People of paradox: An inquiry concerning the origins of American civilization.* Ithaca, NY: Cornell University Press.

Kamuf, P. (1997). *The division/of literature: Or the university in deconstruction.* Chicago, IL: University of Chicago Press.

Kann, R. A. (1980). *A history of the Habsburg empire 1526–1908.* Berkeley, CA: University of California Press.

Kavanaugh, P. (1999). Is psychoanalysis in crisis? It all depends on the premise of your analysis. In R. N. Prince (Ed.), *The death of psychoanalysis: Murder? Suicide? or rumor greatly exaggerated?* (pp. 85–100). Northvale, NJ: Jason Aronson.

Kerouac, J. (1975). *On the road.* New York: Penguin.

Kessner, C. (1994). Introduction. In C. Kessner (Ed.), *The "other" New York Jewish intellectuals* (pp. 1–22). New York: New York University Press.

Kirsner, D. (2000). *Unfree associations: Inside psychoanalytic institutes.* London: Process Press.

Klein, M. (1929/1992). Personification in the play of children. In *Melanie Klein Love, guilt and reparation and other works 1921–1945* (pp. 199–209). London: Karnac Books & The Psycho-Analytic Institute.

Klein, M. (1930/1992). The importance of symbol-formation in the development of the ego. In *Melanie Klein Love, guilt, and reparation and other works 1921–1945* (pp. 219–232). London: Karnac Books & The Psycho-Analytic Institute.

———. (1946/1993). Notes on some schizoid mechanisms. In *Melanie Klein envy and gratitude and other works 1946–1963* (pp. 1–24). London: Karnac Books & The Institute of Psycho-Analysis.

———. (1952/1993). Some theoretical conclusions regarding the emotional life of the infant. In *Melanie Klein envy and gratitude and other works 1946–1963* (pp. 61–93). London: Karnac Books & The Institute of Psycho-Analysis.

———. (1955a/1993). The psycho-analytic play technique: Its history and significance. In *Melanie Klein envy and gratitude and other works 1946–1963* (pp. 122–140). London: Karnac Books & The Psycho-Analytic Institute.

———. (1955b/1993). On identification. In *Melanie Klein envy and gratitude and other works 1946–1963* (pp. 141–175). London: Karnac Books & The Psycho-Analytic Institute.

———. (1957/1993). Envy and gratitude. In *Melanie Klein envy and gratitude and other works 1946–1963* (pp. 176–235). London: Karnac Books & The Psycho-Analytic Institute.

Kohler, L. and Saner, H. (1992). Introduction. In L. Kohler and H. Saner (Eds.), *Hannah Arendt, Karl Jaspers correspondence 1926–1969* (Trans., Robert and Rita Kimber) (pp. vii–xxv). New York: A Harvest Book; Harcourt Brace & Company.

Kohli, W. (1999). Writing the classed body: Is it work yet?. In M. Morris, M.A. Doll, and W.F. Pinar (Eds.), *How we work.* New York: Peter Lang.

Konvitz, M. R. (1994a). Horace M. Kallen. In C. Kessner (Ed.), *The "other" New York Jewish intellectuals* (pp. 144–159). New York: New York University Press.

———. (1994b). Morris Raphael Cohen. In C. Kessner (Ed.), *The "other" New York Jewish intellectuals* (pp. 125–143). New York: New York University Press.

Knowlson, J. (1996). *Damned to fame: The life of Samuel Beckett.* New York: Simon & Schuster.

Kramer, R. (1996). Introduction. Insight and blindness: Visions of Rank. In R. Kramer (Ed.), *A psychology of difference: The American lectures of Otto Rank* (pp. 3–47). Princeton, NJ: Princeton University Press.

Kristeva, J. (2001). *Melanie Klein* (Trans., R. Guberman). New York: Columbia University Press.

Lacan, J. (1955–1956/1993a). On nonsense and the structure of god. In J. A. Miller (Ed.), *The seminars of Jacques Lacan book III: The psychoses 1955–1956* (pp. 117–129). New York: W.W. Norton.

———. (1955–1956/1993b). On the rejection of a primordial signifier. In J. A. Miller (Ed.), *The seminars of Jacques Lacan book III: The psychoses 1955–1956* (pp. 143–157). New York: W.W. Norton.

———. (1955–1956/1993c). The quilting point. In J. A. Miller (Ed.), *The seminars of Jacques Lacan book III: The psychoses 1955–1956* (pp. 258–270). New York: W.W. Norton.

———. (1955–1956/1993d). On primordial signifiers and the lack of one. In J. A. Miller (Ed)., *The seminar of Jacques Lacan book III The psychoses 1955–1956* (Trans., R. Grigg) (pp. 196–205). New York: W.W. Norton.

Laing, R. D. (1960). *The divided self.* New York: Pantheon Books.

Laplanche, J. (1999). *The unconscious and the id* (Trans., L. Thurston and L. Watson). New York: Rebus Press.

Lauter, P. (1996). Strange identities and Jewish politics. In J. Rubin-Dorsky and S. F. Fishkin (Eds.), *People of the book: Thirty scholars reflect on their Jewish identity* (pp. 37–46). Madison, WI: University of Wisconsin Press.

Levinas, E. (1998). *Entre nous: Thinking-of-the-other* (Trans., M. B. Smith and B. Harshau). New York: Columbia University Press.

Lifton, R. J. (1997). Discussion of Martin S. Bergmann's paper. In M. Ostow (Ed.), *Judaism and psychoanalysis* (pp. 152–155). London: Karnac.

Likierman, M. (2001). *Melanie Klein: Her works in context.* New York: Continuum.

Lionells, M. (1999). Thanatos is alive and well and living in psychoanalysis. In R. M. Prince (Ed.), *The death of psychoanalysis: Murder? Suicide? or rumor greatly exaggerated?* (pp. 1–24). Northvale, NJ: Jason Aronson.

Little, F. (1997). *Hyping the Holocaust: Scholars answer Goldhagen.* Merion Station, PA: Merion Westfield Press International.

Long, R. E. (1994). *Ingmar Bergman film and stage.* New York: Harry N. Abrams, Inc., Publishers.

Lothane, Z. (1992). *In defense of Schreber: Soul murder and psychiatry.* Hillsdale, NJ: The Analytic Press.

Macdonald, J. B. (2000). Curriculum theory. In W. F. Pinar (Ed.) *Curriculum studies: The reconceptualization* (pp. 5–13). Troy, NY: Educator's International Press, Inc.

Macedo, D. (2000). Foreword. In R. Sassower, *A Sanctuary of their own: Intellectual refugees in the academy* (pp. ix–xxi). New York: Rowman & Littlefield.

Mahler, M. S. (1979a). On child psychosis and schizophrenia: Autistic and symbiotic psychoses, 1952. In *The selected papers of Margaret S. Mahler: Volume one infantile psychosis and early contributions* (pp. 131–154). New York: Jason Aronson.

Mahler, M. S. (1979b). On symbiotic child psychosis: Genetic, dynamic, and restitutive aspects, 1955. In *The selected papers of Margaret S. Mahler: Volume one infantile psychosis and early contributions* (pp. 109–129). New York: Jason Aronson.

———. (1979c). Perceptual dedifferentiation and psychotic object relations, 1960. In *The selected papers of Margaret S. Mahler: Volume one infantile psychosis and early contributions* (pp. 183–192). New York: Jason Aronson.

Mahony, P. J. (1998). Freud's world of work. In M. S. Roth (Ed.), *Freud: Conflict and culture: Essays on his life, work, and legacy* (pp. 32–40). New York: Alfred A. Knopf.

Maier, C. S. (2002). Mahler's theater: The performative and the political in Central Europe, 1890–1910. In K. Painter (Ed.), *Mahler and his world* (pp. 55–85). Princeton, NJ: Princeton University Press.

Marin, B. (1987). Antisemitism before and after the Holocaust: The Austrian case. In I. Oxaal, M. Pollack and G. Botz (Eds.), *Jews, antisemitism and culture* (pp.216–233). Routledge & Kegan Paul.

Miller, R. (1998). *As if learning mattered.* Ithaca, NY: Cornell University Press.

Mitchell, S. A., and Black, M. J. (1995). *Freud and beyond: A history of modern psychoanalytic thought.* New York: Basic Books.

Moore, M. (2002). *Bowling for Columbine.* A Metro Goldwyn Mayer Film. Dir. Michael Moore.

Morris, M. (2001). *Curriculum and the Holocaust: Competing sites of memory and representation.* Mahwah, NJ: Lawrence Erlbaum and Associates Publishers.

Morson, G. S. (1996). Apologetics and negative apologetics; or, dialogue of a Jewish Slavist. In J. Rubin-Dorsky and S. F. Fishkin (Eds.), *People of the book: Thirty scholars reflect on their Jewish identity* (pp. 78–95). Madison, WI: University of Wisconsin Press.

Munro, P. (1989) Engendering curriculum history. In W. F. Pinar (Ed.), *Curriculum: Toward new identities.* New York: Garland.

Musil, R. (1990a). "Nation" as ideal and reality 1921. In B. Pike and D. S. Luft (Eds.) *Robert Musil Precision and soul: Essays and addresses* (pp. 101–116). Chicago, IL: University of Chicago Press.

———. (1990b). On politics and society: Anschluss with Germany 1919. In B. Pike and D. S. Luft (Eds.), *Robert Musil Precision and Soul: Essays and addresses* (pp. 90–99). Chicago, IL: University of Chicago Press.

———. (1996a). *The Man without qualities vol. 1* (Trans., S. Wilkins). New York: Vintage.

———. (1996b). *The Man without qualities vol. II* (Trans., B. Park). New York: Vintage.

N/A. (1985). Obituaries. Mary Louise Aswell. *The Santa Fe Reporter,* January 9, 1985, p. 6.

Nelson, M. C. (1997). "Marie Coleman Nelson." In A. Molino (Ed.), *Elaborate selves: Reflections and reverie of Christopher Bollas, Michael Eigen, Polly Young-Eisendrath, Samuel & Evelyn Laeuchli, and Marie Coleman Nelson* (pp. 61–104). New York: The Hawthorn Press.

Newfield, C. (1998). Recapturing Academic business. In R. Martin (Ed.), *Chalklines: The politics of work in the managed university* (pp. 69–102). London: Duke University Press.
NRSV. (1989). *The Holy Bible.* The new revised standard version. Iowa Falls, IA: World Bible Publishers.
Ogden, T. H. (1989). *The primitive edge of experience.* London: Karnac.
Ostow, M. (1997a). Discussion of Martin S. Bergmann's paper. In M. Ostow (Ed.), *Judaism and psychoanalysis* (pp. 143–151). London: Karnac.
———. (1997b). Introduction: Judaism and psychoanalysis. In M. Ostow (Ed.), *Judaism and psychoanalysis* (pp. 1–44). London: Karnac.
———. (1997c). The psychological determinants of Jewish identity. In M. Ostow (Ed.), *Judaism and psychoanalysis* (pp. 161–186). London: Karnac.
Painter, K. (2002). (Ed.) *Mahler and his world.* Princeton, NJ: Princeton University Press.
Park, B. (1996). Preface. In Robert Musil's *The Man without qualities from the posthumous papers vol. II* (pp. xi–xvi). New York: Vintage.
Pauley, B. (1992). *From prejudice to persecution: A history of Austrian anti-Semitism.* Chapel Hill, NC: University of North Carolina Press.
Peters, U. H. (1985). *Anna Freud: A life dedicated to children.* London: Weidenfeld & Nicolson.
Phillips, A. (1993). Introduction. In M. Eigen (Ed.), *The electrified tightrope* (xiii–xvi). Northvale, NJ: Jason Aronson.
———. (2000). *Darwin's worms: On life stories and death stories.* New York: Basic Books.
———. (2001). *Promises, promises: Essays on psychoanalysis and literature.* New York: Basic Books.
———. (2002). *Equals.* New York: Basic Books.
Pierce, C. S. (1997/1878). How to make our ideas clear. In L. Menand (Ed.), *Pragmatism: A reader* (pp. 26–51). New York: Vintage.
———. (1997/1904). A definition of pragmatism. In L. Menand (Ed.), *Pragmatism: A reader* (pp. 56–58). New York: Vintage.
Pinar, W. F. (1975/2000). Sanity, madness, and the school. In W. F. Pinar (Ed.), *Curriculum studies: The reconceptualization* (pp. 359–383). Troy, NY: Educator's International Press, Inc.
———.(1994). *Autobiography, politics, and sexuality: Essays in curriculum theory 1972–1992.* New York: Peter Lang.
———. (1998). Introduction. In W. F. Pinar (Ed.), *Curriculum: Toward new identities* (pp. ix–xxxiv). New York: Garland.
———. (2000a). Strange fruit: Race, sex and an autobiographics of alterity. In P. Trifonas (Ed.), *Revolutionary pedagogies: Cultural politics, instituting education, and the discourse of theory* (pp. 30–46). New York: Routledge.
———. (2000b). *The work of intellectuals and the press.* In W. F. Pinar (Ed.), Curriculum studies: The reconceptualization (pp. 422–454). Troy, NY: Educator's International Press, Inc.
———. (2001a). "I am a man" The queer politics of race. *The Journal of Curriculum Theorizing*, 17 (4), pp. 11–42.

Pinar, W. F. (2001b). *The gender of racial politics and violence in America: Lynching, prison rape, and the crisis of masculinity.* New York: Peter Lang.
———. (2004). *What is curriculum theory?* New York: Peter Lang.
Pinar, W. F. and A. E. Pautz. (1998). Construction scars: Autobiographical voice in biography. In C. Kridel (Ed.), *Writing educational biography: Explorations in qualitative research* (pp. 61–72). New York: Garland.
Pinar, W. F., W. Reynolds, P. Slattery, and P. Taubman. (1995). *Understanding Curriculum.* New York: Peter Lang.
Poliakov, L. (1974). *The history of anti-semitism volume four: Suicidal Europe: 1870–1933.* New York: Vanguard.
Pollak, M. (1987). Cultural innovation and social identity in fin-de-siecle Vienna. In I. Oxaal, M. Pollak, and G. Botz (Eds.), *Jews, antisemitism and culture in Vienna.* (pp. 59–74). New York: Routledge & Kegan Paul.
Pulzer, P. (1988). *The rise of political anti-semitism in Germany and Austria.* Cambridge, MA: Harvard University Press.
Rank, O. (1996). *A psychology of difference: The American lectures.* (Ed., R. Kramer). Princeton, NJ: Princeton University Press.
Rapaport, H. (1994). *Between the sign and the gaze.* Ithaca, NY: Cornell University Press.
Readings, B. (1996). *The university in ruins.* Cambridge, MA: Harvard University Press.
Reisner, S. (1999). Freud and psychoanalysis: Into the 21st century. *The Journal of the American Psychoanalytic Association*, 47 (4), pp. 1037–1060.
Rhodes, G. and S. Slaughter. (1998). Academic capitalism: Managed professionals and supply-side higher education. In R. Martin (Ed.), *Chalklines: The politics of work in the managed university* (pp. 33–68). London: Duke University Press.
Ritterband, P. and Wechsler, H. (1994). *Jewish learning in American universities.* The first century. Bloomington, IN: Indiana University Press.
Riviere, J. (1991). *The inner world and Joan Riviere: Collected papers 1920–1958.* London: Karnac.
Roazen, P. (1999). Introduction. In J. B. Agassi (Ed.), *Martin Buber on psychology and psychotherapy: Essays, letters, and dialogue* (pp. xix–xxvi). New York: Syracuse University Press.
Robbins, B. (1990). Introduction: The grounding of intellectuals. In B. Robbins (Ed.), *Intellectuals, aesthetics, politics, academics* (pp. ix–xxvi). Minneapolis, MN: University of Minnesota Press.
———. (1993). *Secular vocations: Intellectuals, professionalism, culture.* New York: Verso.
Rosenblit, M. L. (2001). *Reconstructing a national identity: The Jews of Habsburg Austria during WWI.* New York: Oxford University Press.
Rosenfeld, H. (1965/2000a). Notes on the psycho-analysis of the superego conflict in an acute schizophrenic patient (1952). In H. Rosenfeld (Ed.), *Psychotic states: A psychoanalytic approach* (pp. 63–103). London: Karnac.

———. (1965/2000b). Notes on the psychopathology of confusional states in chronic schizophrenias (1950). In H. Rosenfeld (Ed.), *Psychotic states: A psychoanalytic approach* (pp. 52–62). London: Karnac.

Rosenzweig, F. (1955). Toward a renaissance of Jewish learning. In N. N. Glatzer (Ed.), *Franz Rosenzweig, On Jewish learning* (pp. 55–71). Madison, WI: University of Wisconsin Press.

Roth, J. (1932/1995). *The Radetzsky march* (Trans., J. Neugroschel). New York: Alfred A. Knopf.

———. (1996). *What I saw: Reports from Berlin 1920–1933.* New York: W.W. Norton.

———. (2002a). Strawberries. In *The collected stories of Joseph Roth* (Trans., M. Hoffman) (pp. 138–165). New York: W.W. Norton.

———. (2002b). The place I want to tell you about. In *The collected stories of Joseph Roth* (Trans., M. Hoffman) (pp. 47–53). New York: W.W. Norton.

———. (2002c). This morning, a letter arrived. In *The collected stories of Joseph Roth* (Trans., M. Hoffman) (pp. 166–172). New York: W.W. Norton.

Rucker, N., and K. Lombardi. (1998). *Subject relations: Unconscious experience and relational psychoanalysis.* New York: Routledge.

Rudnytsky, P. L. (2002). *Reading psychoanalysis: Freud, Rank, Ferenczi, Groddeck.* Ithaca, NY: Cornell University Press.

Saltman, K. J. and D. A. Gabbard (Eds.) (2003). *Education as enforcement: The militarization and corporatization of schools.* New York: Routledge Falmer.

Sass, L. (1992). *Madness and modernism: Insanity in the light of modern art, literature, and thought.* Cambridge, MA: Harvard University Press.

Sassower, R. (2000). *A sanctuary of their own: Intellectual refugees in the academy.* New York: Rowman & Littlefield.

Schreber, P. (2000). *Memoirs of my nervous illness* (Trans., I. Macalpine and R. A. Hunter). New York: New York Review of Books.

Schorske, C. E. (1980). *Fin-de-siecle Vienna: Politics and culture.* New York: Vintage.

Schulz, B. (1979). Age of genius. In B. Schulz's *Santorium under the sign of the hourglass.* New York: Vintage.

———. (1999a). "Letter" 1935. In C. Z. Prokopczyk (Ed.), *Bruno Schulz: New documents and interpretations* (p. 13). New York: Peter Lang.

———. (1999b). "Letter 16" 1936. In C. Z. Prokopczyk (Ed.), *Bruno Schulz: New documents and interpretations* (p. 16). New York: Peter Lang.

Schwarz, E. (1999). Mass emigration and intellectual exile from National Socialism. In D. F. Good and R. Wodak (Eds.), *From world war to Waldheim: Culture and politics in Austria and the United States* (pp. 87–108). New York: Berghahn Books.

Scott, J. (1996). Academic freedom as an ethical practice. In L. Menand (Ed.), *The future of academic freedom* (pp. 163–180). Chicago, IL: Chicago University Press.

Sebastian, M. (2000). *Mihail Sebastian, Journal 1935–1944: The Fascist years* (Trans., P. Camiller). Chicago, IL: Ivan R. Dee.

Sedgwick, E. (2003). *Touching, feeling, affect, pedagogy, performativity.* Durham, NC: Duke University Press.

Segal, J. (1995). *Phantasy in everyday life: A psychoanalytic approach to understanding ourselves.* Northvale, NJ: Jason Aronson.

Segel, H. (2000). Notes on symbol formation. In E. B. Spillus (Ed.), *Melanie Klein today: Developments in theory and practice: Volume 1: Mainly theory* (pp. 160–177). New York: Routledge.

Serres, M. (2000). *The troubadour of knowledge* (Trans., S. Faria Glaser with W. Paulsen). Ann Arbor, MI: University of Michigan Press.

Shapiro, H. S. (1999). A life on the fringes—my road to critical pedagogy. In H. S. Shapiro (Ed.), *Strangers in the land: Pedagogy, modernity, and Jewish identity* (pp. 31–29). New York: Peter Lang.

Simon, R. (2000). The touch of the past: The pedagogical significance of a transactional sphere of public memory. In P. Trifonas (Ed.), *Revolutionary Pedagogies: Cultural politics, instituting education, and the discourse of theory* (pp. 61–80). New York: Routledge Falmer.

Simpson, D. (2002). *Situatedness or, why we keep saying where we're coming from.* Durham, NC: Duke University Press.

Sinason, M. (1999). How can you keep your hair on? In P. Williams (Ed.), *Psychosis (madness)* (pp. 44–54). London: Institute of Psychoanalysis.

Sked, A. (2001). *The decline and the fall of the Habsburg empire 1815–1918.* London: Langman/Pearson Education.

Smith, D. (1999). *Pedagon: Interdisciplinary essays in the human sciences, pedagogy, and culture.* New York: Peter Lang.

Steiner, J. (1999). *Psychic retreats: Pathological organizations in psychotic, neurotic and borderline patients.* New York: Routledge.

Stockel, L. (2002). L. Stockel. The Autobiographies. In J. Shandler (Ed.), *Awakening lives: Autobiographies of Jewish youth in Poland before the Holocaust* (pp. 141–196). New Haven, CT: Yale University Press.

Stormer, T. (2002). The Stormer. The Autobiographies. In J. Shandler (Ed.), *Awakening lives: Autobiographies of Jewish youth in Poland before the Holocaust* (pp. 226–262). New Haven, CT: Yale University Press.

Strong, G. V. (1998). *Seedtime for Fascism: The disintegration of Austrian political culture, 1867–1918.* Armonk, NY: M.E. Sharpe.

Sumara, D. (1996). *Private readings in public: Schooling the literary imagination.* New York: Peter Lang.

Sumara, D. and B. Davis. (1999). Inventing a more interesting subject. In M. Morris, M. A. Doll, and W. F. Pinar (Eds.), *How we work.* New York: Peter Lang.

Symington, N. (2002). *A pattern of madness.* London: Karnac.

Taylor, A. J. P. (1948/1976). *The Habsburg monarchy, 1809–1918: A history of the Austrian empire and Austria-Hungary.* Chicago, IL: University of Chicago Press.

Tepa, E. M. (2002). E. M. Tepa. The Autobiographies. In J. Sandler (Ed.), *Awakening lives: Autobiographies of Jewish youth in Poland before the Holocaust* (pp. 275–295). New Haven: Yale University Press.

Thevenin, P. (1989). The search for a lost world. In J. Derrida and P. Thevenin (Eds.), *The secret art of Antonin Artaud* (pp. 1–58). Cambridge, MA: MIT Press.

Timms, E. (1998). Austrian identity in a schizophrenic age: Hilde Spiel and the literary politics of exile and reintegration. In K. R. Luther and P. Pulzer (Eds.), *Austria 1945–95: Fifty years of the second republic* (pp. 47–66). Brookfield, Vermont: Ashgate Publishing Limited.

Trifonas, P. (2003). *Pedagogies of difference: Rethinking education for social change.* Routledge Falmer.

Uhl, H. (1997). The politics of memory: Austria's perception of the Second WW and the National Socialist period. In G. Bischof and An Pelinka (Eds.), *Austrian historical memory and national identity* (pp. 64–94). New Brunswick, NJ: Transaction Publishers.

Updike, J. (1979). Introduction. In B. Schulz's, *Sanatorium under the sign of the hourglass* (pp. xii–xix). New York: Penguin.

Vansant, J. (2001). *Reclaiming heimat: Trauma & mourning in memoirs by Jewish Austrian reemigrees.* Detroit: Wayne State Press.

Villa, D. R. (1996). *Arendt and Heidegger: The fate of the political.* Princeton, NJ: Princeton University Press.

Vergo, P. (2001). *Art in Vienna 1898–1918.* Harrisburg, PA: Phaidon Press.

Wald, A. (1987). *The New York intellectuals: The rise and decline of the anti-Stalinist left from the 1930s to the 1980s.* Chapel Hill, NC: University of North Carolina Press.

Waska, R. T. (2002). *Primitive experiences of loss: Working with the paranoid-schizoid patient.* London: Karnac.

Wat, A. (1988). *My century: The odyssey of a Polish intellectual* (Trans., R. Lourie). New York: New York Review of Books.

Wexler, P. (1996). *Holy sparks: Social theory, education and religion.* New York: St. Martin's Press.

Winnicott, D. W. (1990). The concept of a healthy individual, 1967. In C. Winnicott, R. Shepherd, and M. Davis (Eds.), *Home is where we start from: Essays by a psychoanalyst* (pp. 21–38). New York: W.W. Norton.

———. (1992a). Hallucination and dehallucination. In C. Winnicott, R. Shepherd, and M. Davis (Eds.), *Psychoanalytic explorations* (pp. 39–42). Cambridge, MA: Harvard University Press.

———. (1992b). Introduction to a symposium on the psycho-analytic contribution to the theory of shock therapy. In C. Winnicott, R. Shepherd, and M. Davis (Eds.), *Psychoanalytic explorations* (pp. 525–528). Cambridge, MA: Harvard University Press.

———. (1992c). Physical therapy of mental disorder. In C. Winnicott, R. Shepherd, and M. Davis (Eds.), *Psychoanalytic explorations* (pp. 534–541). Cambridge, MA: Harvard University Press.

Winnicott, D. W. (1992d). The psychology of madness: A contribution from psycho-analysis. In C. Winnicott, R. Shepherd, and M. Davis (Eds.), Psychoanalytic explorations (pp. 119–129). Cambridge, MA: Harvard University Press.

Winter, S. (1999). *Freud and the institution of psychoanalytic knowledge.* Stanford, CA: Stanford University Press.

Wistrich, R. (1999). Introduction: The devil, the Jews, and hatred of the "other." In R. S. Wistrich (Ed.), *Demonizing the other: Antisemitism, racism and xenophobia* (pp. 1–16). Amsterdam, The Netherlands: Harwood Academic Publishers.

Wortham, S. (1999). *Rethinking the university: Leverage and deconstruction.* New York: Manchester University Press; St. Martin's.

Young-Bruehl, Elizabeth. (1982). *Hannah Arendt: For the love of the world.* New Haven, CT: Yale University Press.

———. (1988). *Anna Freud: A biography.* New York: Summit Books.

Zahavi, D. (1999). *Self-awareness and alterity: A phenomenological investigation.* Evanston, IL: Northwestern University Press.

Ziarek, E. P. (2001). *An ethics of dissensus: Postmodernity, feminism, and the politics of radical democracy.* Stanford, CA: Stanford University Press.

Zizek, S. (1993). *The plague of fantasies.* New York: Verso.

Zohn, H. (1943/1964). Introduction. In S. Zweig (Ed.), *The world of yesterday: An autobiography* (pp. v–xii). Lincoln, NE: University of Nebraska Press.

Zuccotti, S. (2000). *The Vatican and the Holocaust in Italy: Under his windows.* New Haven, CT: Yale University Press.

Zweig, S. (1943/1964). *The world of yesterday: An autobiography.* Lincoln, NE: University of Nebraska Press.

Index